CLINICAL BIOETHICS

INTERNATIONAL LIBRARY OF ETHICS, LAW, AND THE NEW MEDICINE

VOLUME 26

The titles published in this series are listed at the end of this volume.

Clinical Bioethics

A Search for the Foundations

Edited by

CORRADO VIAFORA

University of Padua, Italy

Springer

A C.I.P. Catalogue record for this book is available from the Library of Congress.

ISBN-13 978-90-481-6905-4 (PB)
ISBN-13 978-1-4020-3593-7 (e-book)

Published by Springer,
P.O. Box 17, 3300 AA Dordrecht, The Netherlands.

www.springeronline.com

Printed on acid-free paper

Printed in the Netherlands.

TABLE OF CONTENTS

PREFACE

This book, final result of a long term effort who coincides with my scientific coordination of "Ethics and Medicine" Project of Lanza Foundation during its first ten years, collects most theoretically relevant papers of the International Meeting on Clinical Bioethics, held in Padua in 1999, October 19-23.

The Meeting aimed to provide the occasion for comparing theoretical models and formative experiences on Clinical Bioethics, since this was the specific issue focused throught the interdisciplinar seminars organized within the "Ethics and Medicine" Project of Lanza Foundation.

I'm profoundly indebted to the distingueshed colleagues for giving their contribute to the intensive programme of the Meeting both lecturing in plenary sessions on philosophical aspects and analyzing in small groups particular questions concerning Clinical Bioethics.

Most of them moreover have stimulated and enriched the reflection with their experiences in dealing prestigious Centres who have devoted a systematic attention to the foundation of the Clinical Bioethics. I'm particulary grateful to E. Pellegrino and his Center of Clinical Bioethics at the Georgetown University; in particular to R. Dell'Oro; to D. Gracia, director of the Master in Bioethics at the Complutense University in Madrid; to H. Ten Have and his Departement of Ethics, Philosophy and History of Medicine at the Catholic University of Nijmegen; to M. Parker and his Institue for Ethics and Communication in Healthcare Practice at the University of Oxford; to B. Cadoré and P. Boitte and their "Centre d'Ethique Médical" at the Catholic University of Lille; to J. F. Hillhardt and his Zentrum Ethik und Recht in der Medizin at the Albert Ludwigs University of Freibourg; and to P. Drigo, L. Chiandetti and D. Gobber, core-group of Bioethics Committee of Pediatrics Departement of University of Padua.

I hope that the scientific committee of Lanza Foundation will be satisfied in verifying that the specific focus of the Meeting, oriented towards a Clinical Bioethics integrating "internal morality" guided by the goals of clinical practice and "external morality" guided by the wider cultural context, continues to give a stimulating euristic perspective in the international debate.

I want to thank G. Minozzi and R. Pegoraro, past President and President of Lanza Foundation and L. Mariani, president of Scientific committee of Lanza Foundation as well as P. Benciolini and A. Autiero members of the Scientific committee for their suggestions and supports.

I want to thank finally R. Zago, A. Bonanno and K. Calzolari for their precious secreterial support.

Padua, April 2004.

Corrado Viafora

CORRADO VIAFORA

INTRODUCTION

CLINICAL BIOETHICS: A SEARCH FOR THE FOUNDATION BEYOND THE "APPLIED ETHICS" MODEL

Even if from different perspectives, authoritative analyses agree on relating the transformation of medical ethics which occured during the last thirty years to the very transformation undergone by the practice of medicine itself (C. Viafora, 1996).

Sociologists (D. Fox, 1979) have pointed out that medical ethics - understood as the codification of standards of professional conduct - had remained for a long time "segregated" from larger shifts affecting the rest of culture. That "segregation" was justified by the predominant conviction that the codification of medical conduct is the exclusive competence of physicians. Indeed, one can say that medical ethics was the ethics of physicians (D. Von Engelhardt, 1995).

Although for Southern European cultures the emphasis is still on professional deontology, in Northern European as well as in North American countries medical ethics has been progressively framed - at least since the 1960s - within a larger context defined by the word "bioethics".

The first meaning of this shift from traditional medical ethics to bioethics is the new attention given to ethical issues which fall beyond the boundaries of the physician-patient relationship. These issues include, among others, the relation between health care professionals, clinical research and experimentation on human subjects and criteria for the just allocation and distribution of medical resources.

The shift, however, is not confined to the emergence of new topics. It is also of a more formal nature for it includes a new way of approaching ethical quandaries. Bioethics can be seen historically as the inevitable result of the application to the biomedical field of those principles such as pluralism, moral autonomy, democracy and human rights which have been influencing the culture of Western countries for at least two centuries. In this perspective bioethics signals the end of the segregation of medical ethics and its initiation into the era of modernity (D. Gracia, 1989; 2001).

Given that this is the meaning of the transformation from medical ethics to bioethics, it is not difficult to understand why bioethics has been fundamentally understood as a form of "applied ethics." Once they have entered into the process of modernity, medical practice and health care become cultural phenomena embodying the rules of the larger context to which they belong. Consequently, bioethics can very well be conceived as

> "the application of theories, principles, and general ethical rules to the particular problems of therapeutic practice and of biomedical and biological research." (T. Beauchamp and J. Childress, 1983).

This model of bioethics seems to be defined by the search for an *external* point of view, i.e., one which can critically look at medical practice only because it remains in an outside position, so to speak. Yet, the emphasis on such an external perspective is not without consequences: to start with, the ethical reflection ends up losing contact with the clinical context and the medical practice (B. Cadoré, 1997). Moreover, the physician-patient relationship, rather than being viewed for what it is in itself, becomes prey of contractualist models of interpretation, that is, models which being foreign to the particularity of the physician-patient relationship, cannot account for its specific intentionality (W. Reich, 1993; E. Pellegrino and D. Thomasma, 1988). What is problematic in this situation, however, is not only the progressive alienation of ethical reflection from concrete clinical experience. Given the pluralistic nature of our society, it has become difficult to find an external perspective shared by everyone that could function as an impartial point of view. Such a perspective would merely consist on the adoption of a "neutral moral language." (T. Engelhardt, 1986). For its critics, the emphasis on a presumed neutral morality carries, in reality, the mask of a very specific, and by no means, neutral ideology, namely, the "ideology of pluralism" (C.S. Campbell, 1996). Such a position mistakenly pretends to separate the self from "the sources of his identity" (Ch. Taylor, 1989) and can only result in a "minimalistic" ethics completely unprepared to deal with the ethical challenges of medicine and health care (D. Callahan, 1980).

But is it possible at all to keep together *proximity* in clinical practice with critical *distance*?

European approaches to clinical bioethics tend to solve the problem by searching for a theoretical structure that tries to interpret the ends of medicine *from within,* yet assures at the same time the conditions for critical distance. To put it in a somewhat technical way, the problem is to integrate "internal" and "external" morality: the former flows directly from an intepretation of the ends of medicine, the latter conveys the juridical framework and the culture which underlies the practice of medicine (Henk Ten Have, 1995).

Bioethics reflects the polarization of the broader ethical debate between an ethics centered on the notion of *right* and an ethics of the *good.* The polarization identifies, respectively, the libertarian and the communitarian positions. According to the first, the debate, both in ethics and bioethics, must refrain from referring to a particular notion of the good life and simply address the issue of how it is possible to assure the peaceable co-existence of different positions. The other position views the discussion on the many notions of "the good life" as central to the debate: because different conceptions of "the good life" always underlies a certain biomedical practice. It is an illusion to think that it is possible to focus on what is right on the condition that all the notions of the good be neutralized.

Perhaps the polarization just mentioned can be recast in terms of a counterposition between a *procedural* approach in which the determination of what is right is reached without reference to a particular moral tradition, and a *substantive* approach, which denounces the former as fundamentally abstract and emphasizes the social integration of the individual and the important role of moral traditions (McIntyre, Sandel, Ch. Taylor, S. Hauerwas).

The theoretical structure of medical ethics should be re-defined by trying to integrate internal and external morality. Indeed, an important contribution to the debate may come from a reflection that tries to hold together both rational procedures based upon principles and ethical convictions inspired by particular traditions. In various guises, this argument leads toward a confrontation between the universalist claim attached to procedural rules and the recognition of positive values belonging to the historical communitarian contexts of the realization of these same rules. In this sense, the universalist claim and the contextualist claim have to be maintained together, entrusting to the practical wisdom of "moral judgment in situation" to surmount *from within* the possible antinomy of the two (P. Ricoeur, 1993).

The reference to the specific function of moral judgment and to the importance of practical wisdom as central components in the integration of external and internal morality is not entirely new, but rather carries on a central trait of Western ethical tradition with its emphasis on the essential role of deliberation. In today's clinical context the traditional notion of deliberation must be perfected and integrated by a precise methodology of clinical judgment. The more disciplines and cultural perspectives get involved in the ethical analysis of clinical quandaries, the more emerges the importance of deliberation reached through interdisciplinary dialogue, and the more clearly appears the need for both, a methodology of clinical judgment (D. Gracia, 1991; B. Cadoré, 1995; A. Autiero, 1995; C. Viafora, 1999) and the education of health professionals (P. Poletti, 1995; S. Bastianel, 1997).

These premises can suffice in sketching out the scope of this book. The title itself suggests a model of *clinical bioethics* capable of bringing together two attitudes, i.e., a critical view of health care practice and adherence to the intrinsic ends of medicine. If the term "bioethics" conveys the meaning of the former attitude, the adjective "clinical" expresses the fundamental intention of the latter.

With this concept of clinical bioethics in mind, the book pursues the following aims: (i) to present and confront different foundational approaches to clinical bioethics (ii) to identify the conditions under which it becomes possible to integrate internal and external morality.

After a comprehensive introduction, with particular reference to the ethical crisis in the post-modern age (D.Thomasma), the first section presents two approaches which give an alternative interpretation of the shift from Medical Ethics to Bioethics: the approach proposed by E. Pellegrino (an exemplary internalist approach) focusing on redefining the "good of the patient", which is the basic principle of the Hippocratic tradition and the approach proposed by D. Gracia, (an exemplary externalist approach) focusing on the foundation of Medical Ethics in the perspective of the democratic evolution of modem society.

> "Bioethics - D. Gracia affirms - is a process of deliberation about the individual and collective ends of human life. Consequently, it can not be constrained to the limits of hospitals and schools of medicine. The goals of human life are primarily social and political. And because the ends of medicine are derived from these goals, it is necessary to conclude that bioethics is engaged inevitably in matters that occur out of the health care professions".

Then, we take into account three different contributions, centered on the introduction of new approaches to Clinical Bioethics, which propose specific perspectives in integrating internal and external morality: *a communitarian approach* based on the dialectic interaction between anthropology and ethics (H. Ten Have); *a hermeneutical approach* based on the consideration of the action as a subject to which apply the main rules of the interpretation (B. Cadoré); and *a deliberative approach* based on the dialogical relationship between individual and community (M. Parker).

The second section of the book aims at showing the conditions required to integrate internal and external morality: this approach to the clinical bioethics has been suggested by H.Ten Have (H.Ten Have, 1995, 2001), it justifies the relevance given in the structure of the book to his contribution. In programmatic attempts and debates pursuing the aim of connetting internal and external morality, H.Ten Have identifies the follow steps. The first step is to examine the internal standards and normes that govern the medical practice in the different care contexts in order to obtain a better understanding of the internal morality in terms of good clinical practice. The second step is to analyse and interpret the external morality in order to understanding the cultural contexts regarding health, disease, disability, dying, illness, prevention and health care. These steps requires making use of the results of specific empirical investigations. The third step is to create a new theoretical approach to health care institution, since it is the concrete context in which internal and external morality interact. The modern health care system is a complex network of practices based on different values and using different methods. Focusing the neo-aristotelian notion of practice (A. MacJntyre, 1982) as theoretical starting-point, H.Ten Have suggests as heuristic instrument to frame the ethical problems of the modern health care system the distinction proposed by U.Jensen: the *disease* oriented practice; the *profession* oriented practice and the *community* oriented practice (U.Jensen, 1987). The fourth step is to develop, in the perspective of this community oriented approach, a new conception of clinical bioethics aimed to integrate the normative approach with the ermeneutical approach. According to H.Ten Have, only this integration ("interpretive bioethics") can illuminate and clarifies the complex interaction between the internal and external morality of health care practice.

In this perspective, two series of reflections are offered: (i) the first one concerns the redefinition of the goals of medicine both in relation to the subjetivisation of health (R. Mordacci, P. Zatti); and in relation to the rationalisation of the health care system (P. Vineis, H. Jochemsen). If the reference to the current perception of the health and the evolution of the health-care system leads us to compare internal and external morality in relation to the goals of human life, the last part of the book is focused on ethical judgment, with particular reference to its epistemological statute (R. Dell'Oro), and to the institutional context of clinical practice (P. Boitte, C. Viafora).

A CRITICAL EXAMINATION OF THE "INTERNAL MORALITY" OF MEDICAL PRACTICE: A SYNTETICAL ACCOUNT OF THE CURRENT DEBATE

In recent times the debate on redefining the goals of medicine has attracted a considerable attention. (M. J. Hanson, D. Callahan, 1999). Some problematic issues have contributed to analyse in depth the normative meaning of the medical practice: physician - assisted death, managed care, judgments of medical futility.

An interesting issue of "The Journal Medicine and Philosophy" edited by R. Veatch and F. Miller (6, 2001) has esamined in particular the concept of the internal morality of medicine with the aim of clarifing its meaning and value. This issue, conceived as lively symposium, proposes two papers "pro" thesis of internal morality, and two papers "contra". In the first paper "pro", E. Pellegrino (E. Pellegrino, 2001) in a revised version of a paper delivered at International Meeting on Clinical Bioethics held in Padua (october, 1999), now published in this volume, offers an explication of the internal morality as graunded on the phenomena of medicine, with particular reference to the nature of the clinical encounter between physician and patient. In the second paper "pro", F. Miller and H. Brody (F. Miller-H., Brody, 2001) develop a critical examination of the conception of the "internalist" perspective advocated by E. Pellegrino toward an understanding of the internal morality notion in an "evolutionary perspective". That evolution for them takes place in dialogue with the human history and culture.

The two papers "contra" rejects the internal morality concept on the basis of the following critical considerations: (i) we dont need an internal morality attributable to medicine "qua" medicine, we can resolve the moral problems in clinical practice by the systematic "specification" of the external, i.e., common morality (T. Beauchamp, 2001); (ii) an internal morality for the medical practice is impossible, because it is impossible to know the ends of the medicine without knowing the ends or goals of human living (R.Veatch, 2001)

In the last essay of the *Journal,* J.Arras, as critical commentator, concludes that, even if is very difficult to discern a set of moral norms internal to medical practice, neverthless the notion of internal morality can be defended as giving a general orientation toward the "virtues necessary" to practice the medical profession (J.Arras, 2001).

The solution proposed by J.Arras recognizes that there are good reasons in the work of both the defenders and the critics of the internal morality thesis. Instead of viewing internal medical morality as a guide to the resolution of substantive moral problems, Arras proposes to advocate internalism by assigning it a more modest function, for witch the proper function of this morality is not to solve problems, but rather to give physicians an identity as professionals, rather than a self-interested tradespeople, and a basic education in some key medical virtues (courage, compassion, truthfulness, etc) (J.Arras, 2001).

In conclusion of this syntetical account of the debate about the internal morality of medicine exemplarly proposed by the *Journal*, my opinion is that there are at least two reasons which prevent from an adeguate understanding of the attempt to an integration between internal and external morality: the first reason is the functionalistic conception of the medical profession; the second reason is the enphasis in the normative dimension of ethics and bioethics.

TOWARD A REINTERPRETATION OF THE INTERNALIST PERSPECTIVE ON THE BASIS OF THE PRACTICAL NATURE OF THE MEDICAL RATIONALITY

Theoretical presuppositions

An adequate understanding of this attempt can be finded on the stimulating reinterpretation of the internalist perspective of E.Pellegrino, one of the early proponends of the internal morality concept, elaborated by R.Dell'Oro (R.Dell'Oro, 2003).

The start-point of this reinterpretation is a clear identification of the limits both of "essentialistic" and "functionalistic" approaches to clinical bioethics. According to Roberto Dell'Oro, the essentialistic approach derives the internal ends from the peculiar nature of medicine, on the basis of a phenomenological analysis of its constitutive components. In this approach, medicine cannot receive some "goals" which are different from its constitutive "ends". Medicine will be always and everywhere a healing act oriented to the fact of illness; a relation based on trust routed in the clinical encounter and having as *telos* the patient's good. The reason for considering this approach as essentialistic is based on its presumption of identify the essential nature of medicine "a priori", outside of history. In contraposition to this essentialistic approach, the functionalistic approach considers the ends of medicine as derived from a completely contractualistic process. Therefore medicine don't possesses any internal end and the criteria of medical ethics don't are different in nature from the general principles of "public ethics" or "common morality". They are only a "regional" application and specification of these principles, without any reference to the phenomenological specificity of the medical practice. The reason for considering this approach as functionalistic is referred by R.Dell'Oro to the presuppositions that: (i) the ends of medicine must be considered as a contractualistic determinatiion of the individual's ends; (ii) the interpersonal dimension of the clinical encounter is substituted with a merely transational relationship, entirely determined by the particular socio-cultural context.

According to R.Dell'Oro, the solution of the contraposition between the essentialistic and functionalistic approach can be given by focusing the practical nature of the medical rationality. Just because the medical rationality is practical in nature, i.e., it refers to the action and therefore to the free determination of the human will, it is clear, on the one hand, that the ends of medicine must be defined in the historical way; on the other hand, it is clear that the even necessary contextualization of the medical practice don't can deny to this practice the space of its relative autonomy in determining its ends.

In conclusion, the consideration of the practical nature of the medical rationality can mean to interpret the ends of the medicine facing to tension between anthropological presupposition and phenomenologycal essentiality. In this reinterpretation, the internal morality of medicine becomes an *open system*, in which the internal ends are internal not because exclusively determined by the medical profession, but because they inspire the determination of the patient's good, which is the specific end of the medicine "qua" medicine. In a more radical sense, the good that medicine must make possible is not "internal", nor "external", because the concrete determination of the patient's good emerges within a relational process based on the dialogue and communication.

Methodological suggestions

An adequate integration between internal and external morality requires a more robust conceptual frame in order to clarify the semantic of both internal and external morality. It also requires a methodology of ethical jugment. The following considerations aim to suggest an argumentative scheme in order to integrate within the analysis of clinical practice internal and external morality (Viafora, C, 1999).

Before illustrating the different areas of this scheme and their specific functions, an important meta-ethical premise must be focused. The priority explicitably given in this argumentative scheme to internal ends of medical practice implies approaching clinical bioethics and its system of argumentation from a *teleological* point of view, itself grounded in the Aristotelian tradition. The opposite *deontological* approach inspired by Kant is thereby put in a secondary position. Yet, this don't completary clarify all the presupposition adopted. Someone, like A. MacIntyre, tends to play the two approaches one again the other. On the other hand, P. Ricoeur interprets their relation as dialectical. Pratical jugment derives from both approaches and ultimately the argumentational structure is grounded on the reciprocity of deontological and teleological moments.

In this perspective, the starting-point of a systematic argumentation in clinical bioethics is the *moral experience* internal to clinical practice. Such experience refers to the ends witch constitute and define the practice itself, providing a guiding paradigm, that may be identified in the following moral issues: protection of life, promotion of the patient health, respect of his personal dignity and fairnes in the allocation of community resources. Even if this ethical paradigm cannot provide a normative scheme, neverthless it is a source of moral creativity and personal motivation.

The application of this ethical paradigm always takes place within *a particular clinical context*: the obstetric context, the pediatric context, the geriatric context, the oncological context, the intensive therapy context, etc.. The specific ends defining each clinical context will provide the first bases for articulating the ethical paradigm of the clinical practice. The forms of respect for the personal dignity of the patient vary according to the context in witch it is played out. The ethical reasoning has to articulate the meaning of respect for each situation, recognizing for the intrinsic possibilities of good it possesses. Referring to the general category of respect without taking in to account the particularity of the respect for which it should make sense, will lead to abstract solution.

Right next to the area comprising the ends of the medicine is the area of the *virtues*. If the ends internal to the practice of the medicine may play an important function in setting up an horizon of ideals, and the clinical context, in turn, provides the situation background against which the different moral questions are to be framed, yet, these two criteria are valid only when they are appropriated at a personal level and when they form the personality of the moral subject. This is, indeed, the specific function of virtues. Thanks this process of personal appropriation, virtues come to define the particular sensibility which directs the moral agent, influencing his global motivational structure.

The importance of different clinical context and corresponding sensibilities in interpeting the ends internal to clinical practice can be fully appreciated within a larger *cultural context*, where are "the sources of the self" (C. Taylor). The frame of values which substantiate a particulare culture of life represents the hermeneutical background for understanding clinical practice. Moreover, such a background structures the perception of the ethical problems and contributes to the determination of their general features. The meaning of life, of suffering and death, of illness and health are always embedded within a particular cultural texture: a symbolic matrix which – implicitly or explicitly – fashions the interpretation of the clinical practice and its internal ends. The awareness of such a cultural frame is important for the ethical analysis. Since the ends of clinical practice are seen within the general precomprehension, it becames easier to identify the particular level of an ethical conflict. It is, first and foremost, one of anthropological presuppositions rather than different normative solutions.

Taking the moral experience internal to clinical practice as starting point of the ethical analysis, does not thereby imply abandoning any influence to normative principles. The funcion of principles within the ethical analysis is to regulate and to order the ends of internal morality, when this ethical claims conflict with one another for the complexity of the particular cases. Ultimatly this regulating function of the principles consists in a formal strategy where by the application of the ends of the moral practice to particular cases becomes possible. As a *clinical* strategy, its starting point will be the clinical encounter. As an *ethical* strategy, the application of the norm to the particular cases must obey to the principles of universalization. If this strategy gives the formal condition for their application, the specific contents of the bioethical principles, according to the integration of internal and external morality, can be reinterpreted by the following sequence: first, the principle of beneficence; second, the principle of autonomy and third the principle of justice.

The first principle to rank the ends internal to clinical practice will be the *principle of beneficience*. Its overal goal is the good of the patient, whether at a diagnostic or therapeutic level, for the phase of prevention or rehabilitation. The very health care profession implies a public promise to act for the patient's good. According to this promise, a health care professional acts ethically if she takes the patient's good as a general end. In general terms, the principle can be formulated like this: *act in such a way that the consequences of your intervention will be for the patient's good.*

Negatively, the principle imposes the obligation to do no harm to the patient.

Positively, its promotes the patient's good by assessing the proportion between the benefits and the risks of any medical intervention. The emergence of a movement for the patient's emancipation and the development of modern medicine in its diagnostic, therapeutic and rehabilitation power, have triggered conflicts and levelled objections to the principle. Indeed, the possibility of conflicts between physician and patient is increased by the need to take into account the patient's subjectivity and the wealth of options available for treatment.

In case of persistent conflict between a suggested medical treatment and the patient's will, the last word ought to be left to the patient. The principle of autonomy prevails because the patient remains always responsible for his life and decisions. In general terms, the principle can be formulated as follow: *act in such a way as to respect the patient in his personal dignity and in his right to responsibly decide whether to accept or to refuse a suggested treatment.*

Negatively, the principle of respect implies refraining from interference and intrusion. Positively, respect for autonomy demands that the physician adequately inform the patient and involve him in decision making process. Respect for autonomy goes hand in hand with the need to assess what the patient really wants. Of course, upon establishing its authenticity, the patient's will must be respected.

If the principle of beneficence directs the actions of the health care professional toward the medical end, and autonomy deals with the particular responability of the patient, it is the principle of justice that represents and articulates the needs of society. The increasing costs of health care need to be limited. Yet, cost containment cannot be a primary concern of the physician, especially when this affects the personal relation with the patient. Setting limits is, rather, a political problem since the responsibility for the common good is entrusted to political power. At the level of political intervention, the moral bottom line is defined by the principle states the following: *in the allocation of health care resources act in such a way that privileges and burdens be distributed fairly, i.e., without discrimination in the treatment of persons, unless this is required and justified for the advantage of those who are most in need.*

The specific contribution of clinical ethics to the problem of allocation is to ensure an effective administration of the community's resources. The principle of justice challenges the medical profession in renewing menagement of resources by the medical profession. The latter, in turn, is preserved against the intrusion of political decisions affecting cost containment from tempering in any way wiht the intrinsic ends of clinical practice. The challenge to clinical practice today is being levelled from both sides: the perspective of *distributive justice* and of *commutative justice* which can be expressed in the following way: *act in such a way as to grant to each person the some degree of respect and consideration.*

The urgency to retrieve such a notion of justice derives from the development of new diagnostic and therapeutic treatments. The degree of application of such new treatments extends to every sphere of life and unmasks the most intimate aspects of each person's life. In areas like reproduction, death and dying, genetics, the dangerous possibility of intruding, selecting, and ultimately, discriminating against individuals always exists. Bioethics cannot count upon the legacy of the modern philosophycal tradition in rearticulating the meaning of commutative justice. The language of rights borrowed from this tradition has, indeed, led to the recognition that each person deserves the same respect and consideration. Yet, it hasn't provided the answer to the question concerning "who is the other" entitled to respect and consideration. More specifically, the main flaw of the modern philosophycal tradition can be found in its inability to grasp the meaning of the biological dimension in the subject's constitution. This inability has proven full of consequences for ethics as well. Suspending the biological dimension from ethical consideration has prepared the ground for possible discrimination. If, in fact, the biological dimension cannot exhaust all aspects of the human being, it represents, nevertheless, the necessary condition to be taken into account in order to protect each person. An attitude of protection toward physical life does not entail a biologist interpretation, it simply provides a shield against an opposite proclivity toward selection. Indeed, anyone can claim the right to be treated as "the other" and to be included in the moral community on account of his or her body.

The deontological sequence of beneficience, autonomy and justice has been unfolded in its meaning – to order the ends intrinsic to clinical practice – and it has, therefore, come to its final limit. Limit in double sense. First, the principle of justice draws the line between what is negotiable and what is not. In the latter sense, the principle of justice imposes upon the individual's coscience duties which are absolute. Kant calls them perfect duties: the individual's autonomy does not dispose of them and for this reason cannot refuse to obey. Second, the limit is also a limit of content. In dealing with issues of distributive and commutative justice, clinical ethics extends beyond its specific competence and steps, respectively, into the field of politics and law.

Corrado Viafora, Professor of Moral Philosophy and Bioethics, Faculty of Educational Sciences and Faculty of Medicine, University of Padua, Italy.

REFERENCES

Arras, J., (2001), *A method in the search of a purpose: the internal morality of medicine*, in "The Journal of Medicine and Philosophy", 6, (2001), pp. 643 - 662.

Autiero A. (1997), *Elaborazione sistematica dell'argomentare etico, in* AA-VV., *Verità e Veracità,* Atti del XVI Congresso Nazionale, ATISM, Napoli , pp. 117-139.

Bastianel S. (1997), *Quale formazione per la bioetica,* in "Credere Oggi", n. 4, pp. 92-103.

Beauchamp T., Childress J. (1983), *Principles of biomedical ethics*, Oxford University Press, New York, II ed., pp. IX- X.

Beauchamp, T., (2001), *Internal and external standards for medical morality*, in "The Journal of Medicine and Philosophy", 6, pp. 601 - 619.

Cadoré B. (1997), *L'éthique clinique comme philospophie contextuelle,* Fides ed. , Montreal

Cadoré B. (1995), *La méthodologie en éthique clinique,* in "Forum Ethique", *Quelle éthique pratique en médicine?,* Editions E.C.C.E., Bruxelles, pp. 19-43.

Callahan D. (1981), *Minimalist ethics,* in Caplan A., Callahan D. (eds.), *Ethics in hard times,* Plenum Press, New York, pp. 261-281.

Campbell C.S. (1996), *Bioethics and the spirit of secularism,* in E.E. Shelp (ed.), *Secular bioethics in theological perspectives,* Kluwer Academic Publ., Dordrecht, pp. 3-18.

Dell'oro, R., (2003), *La bioetica clinica come etica applicata. Problemi teorici della consulenza etica nell'istituzione sanitaria,* in Picozzi, M. Tavani, M. Cattorini, P. (a cura di), *Verso una professionalizzazione del bioeticista. Analisi teorica e ricadute pratiche*, Giuffrè editore, Milano.

Engelhardt T. (1996), *The foundations of bioethics,* Oxford University Press, New York.

Engelhardt Von D. (1989), *Ethik in Altag der Medizin. Spektrum der medizinischen Disciplinen,* Springer - Verlag, Harberg,.

Fox D. (1979), *The segregation of medical ethics: a problem on modern intellectual history, in* "The Journal of Medicine and Philosophy", pp. 81-97.

Gracia D. (1989), *Fundamentos de bioética*, Eudema Madrid 1989.

Gracia D. (1991), *Procedimientos de decisiòn en ética clinica*, Eudema, Madrid.

Gracia D., (2001), *History of medical ethics*, in Ten Have, H. Gordijn, B. (eds), *Bioethics in the european perspective*, Kluwer Academic Publishers, Dordrecht, 2001, pp. 17-50.

Jensen, U.J., (1987), *Practice and progress. A theory for the modern health care system.* Blackwell Scientific Publications, Oxford.

Hanson, M. J., Callahan, D., *The goals of medicine : the forgotten issues in Health Care Reform*, Washington D. C., Georgetown University Press, 1999.

The internal morality of medicine, (2001), issue edited by Veatch R., and Miller F., in "The Journal of Medicine and Philosophy", 6.

Ladrière J. (1982), *L'interdisciplinarité. Réflexion épistemologique et philosophique*, paper quoted by B. Cadoré, *L'expérience bioétique de la responsabilité*, Artel-Fides, Louvain- la- Neuve,1994, p.22.

MacIntyre, A.C., (1981), *After Virtue*, University of Notre Dame Press, Notre Dame.

Miller, F., Brody, H., (2001), *The internal morality of medicine: an evolutionary perspective*, in "The Journal of Medicine and Philosophy", 6, pp. 581 - 599.

Pellegrino E., Thomasma D. (1988), *For the patient's good: the restauration of beneficience in health care*, Oxford University Press, New York.

Pellegrino, E., (2001), *The internal morality of clinical medicine: a paradigm for the ethics of the helping and a healing professions*, in "The Journal of Medicine and Philosophy", 6, pp. 559 – 579.

Reich W. (1993), *Alle origini dell'etica medica: mito del contratto o mito di cura?* in Cattorini P., Mordacci, R. (a cura di), *Modelli di medicina. Crisi e attualità dell'idea di professione*, pp. 35-59, Europa Scienze Umane ed., Milano.

Ricoeur P., (1990), *Soi-même comme un autre*, Seuil, Paris.

Ricoeur P., (1996), *Les trois niveaux du jugement médical*, in "Esprit", 12, pp. 21-33.

Taylor Ch., (1989), *Sources of the self. The making of modern identity*, Cambridge University Press, Cambridge.

Ten Have H. (1995), *Approcci europei all'etica nella medicina clinica,* in Viafora C. (a cura di),*Comitati etici. Una proposta bioetica per il mondo sanitario,* Fondazione Lanza-Gregoriana ed., Padova.

Ten Have H. (2001), *Theoretical models and approaches to ethics*, in Ten Have, H. – Gordijn, B. (eds), *Bioethics in the european perspective*, Kluwer Academic Publishers, Dordrecht, 2001, pp. 51- 82.

Veatch, R., (2001), *The impossibility of the morality internal to medicine*, in "The Journal of Medicine and Philosophy", 6, pp. 621 - 642.

Viafora C., (1996), *Bioethics: a history,* International Scholars Publications, Bethesda, 1996.

Viafora C. (1999), *Toward a methodology for the ethical analysis of clinical practice,* in "Medicine, Health care and Philosophy. An European Journal", 2 (3), pp. 283-297.

PART 1

CLINICAL BIOETHICS: COMPARING THEORETICAL MODELS

DAVID C. THOMASMA †

CLINICAL BIOETHICS IN A POST MODERN AGE

The second half of the 20th Century was a good time for applied ethics. Prior to that era, ethics had become concerned about meta-ethical questions like determining the good, and methodological and epistemological concerns about to what ethical statements actually referred[1]. While these debates were occurring, the world around us changed dramatically. The ready alliance of technology with armaments, and the service by onceindependent disciplines of the interests of the state, led to catastrophic results for the human community worldwide[2]. Controlling science and technology and aiming it at good human ends, especially after World War II, became such an enormous problem that ethics could no longer pour over textual matters in isolation from the world impinging on it[3]. It was summoned to the service of humanity in the form of national and international commissions. Practical, or applied, ethics, was an idea whose time had come.Since that time, both popular and professional interests have zeroed in on ethical dimensions of medicine, nursing, allied health, law, business, engineering, politics, the military, education, and many other professions[4]. It has become so much a part of modern life that even media interviewers instinctively ask their subjects to comment on the ethical implications of new discoveries. Widely-accepted standards have emerged from such media attention, study, and public dialogue. In medical ethics, for example, such standards range from agreements about allocating scarce resources like livers, the requirement of informed consent, to the rights of patients to determine their own treatment. Comparable standards have been set in the other fields mentioned. Current interest lies in broadening practices and policies in the international community[5].

Behind these advances in public moral dialogue lies a simple fact. The ethical dilemmas faced by individuals and societies are difficult, personal, emotional, and very much in the present. Faced with the immediacy of the problem, individuals want some answers that are defensible. Even if one's instincts were to apply the ethical theories from the past in such an environment of immediacy, this would no longer be an academic task. Whether families ever read Aristotle or Kant does not matter. In dialogue with professionals they have to formulate their own values and express them as honorable and worth upholding. The resulting discussion is the warp and woof of ethics. Thus the move to everyday problems in ethics helped ordinary individuals enter the moral dialogue by expressing their experiences that "welled up" from their encounters with new technological practices.

By engaging in the everyday ethical dilemmas caused by advancing technology, and by honoring the individuals or units of society (such as families, or doctors) who articulate moral problems, ethics moved out of its academic gown into the hospital wards, nursing homes, pharmacy practices, scientific research centers, political arenas, and government institutions. This move has been spectacularly successful, so

C. Viafora (Ed.), Clinical Bioethics. A Search for the Foundations, 3-20.

much so that graduate programs abound at the master's and doctorate level that highlight or concentrate on medical, nursing, legal, and other applied ethics disciplines. Some ethicists are viewed as “experts" and are brought to the problem area, like the bedside in the case of medical ethics, to assist in decision making and to make recommendations. Ethicists have been expert witnesses in court trials, even taking the stand against the views of one another[6], giving rise to the very question of the possibility of having philosophical "expertise" that can avoid ethical sophism.

In this paper, I will focus on one field of applied ethics, namely, clinical medical ethics. In this analysis I borrow a distinction from Corrado Viafora, that clinical bioethics is a bridge between clinical ethics (a branch of clinical judgment in medical practice) and applied ethics generally[7]. In doing so I wish to focus on the possible foundations of clinical ethics as just described, the foundations becoming the groundwork of clinical bioethics and philosophy of medicine. Clinical bioethics must focus on the experience of patients and health professionals in practice as a legitimate moral realm, the realm of clinical practice.

Interesting sequellae resulted from the move of ethics to the practical realm. Let us look at a few of them before further defining the focus of this paper.

CONSEQUENCES OF PRACTICAL CONCERNS IN ETHICS

The first problem that happened to ethics in the practical sphere was that theory did not seem to matter any longer. Persons of honorable intentions, despite widely divergent backgrounds, training, and culture, found themselves in agreement about strategies. This point first surfaced in reviews of Beauchamp and Childress' pathbreaking text that set the standards for bioethics and in subsequent editions continues to do so. Beauchamp, a utilitarian, and Childress, a deontologist, developed a theory derived from W. D. Ross that formulated agreement on middle-level principles of autonomy, beneficence, non-maleficence, and justice, despite their different allegiances to ethical theory. These are seen, following Ross, as *prima facie* principles, which means that they are able to be ordered hierarchically only when one principle could be prioritized over another in a specific case. Later editions of their book expanded on the authors' suggestions for a strategy and methodology regarding how to accomplish such prioritizations.

Reading the book, scholars began to ask whether one's theory mattered at all, since resolving ethical dilemmas in medicine and in clinical ethics seemed to depend instead on balancing *prima facie* principles that could be derived from any number of ethical theories among which there is little agreement. What is the good of theory then?

The second problem emerged from the first. If theory did not seem to matter, why bother with it at all? The oddity about ethics is that persons agreeing in principle can reach different conclusions. By contrast, persons disagreeing in principle could reach amicable unanimity about practical strategies or public policy. The reason for this lies in the fact that decision making requires a comfort level, since every decision engages a balancing of values. Hebert Simon, an economist known for his work on decision making four decades ago coined the term "satisficing," to explain that decisions are typically compromises in which the benefits to be earned by careful and rational consideration are weighed against to need to come to a decision. This leads to decisions about practical matters that are somewhere in between optimal and merely satisfactory, what Simon called "bounded rationality"[8]. His work suggests that there is much more to clinical ethics decision making than rational analysis of principles coupled with intuitive or commonly-shared cultural values, something more than logical consistency or optimizing behavior. But what is that extra dimension?

The third consequence of note was already mentioned. If one accepts the primary role of those most affected by decisions made by others, these individuals become “stakeholders" in that decision. Generally there is an assumption, correct I think that the people closest to the moral dilemma have the most to gain or lose in an ethical analysis. They are therefore not only valued (from the perspective of moral analysis) for their articulation of the moral problem, but also as participants in the resulting moral discourse that should be part of medical practice. This perception demands that clinical bioethics be participatory in a directly democratic way, a way that a more elitist academic discipline of ethics is not, nor indeed, could not be.

The fourth sequellae is that the major focus of clinical bioethics must be procedural. Following Gracia, Viafora notes that there are two reasons for this. The first is that the complexity of modern health care requires serious attention from the point of view of many more interests than in the past. Setting procedures for adequate and just participation of all stakeholders is required. Second, many decisions are made from too remote a perspective for those with an interest in the results[9]. The worry that proceduralism would be compressed down to strategic ethics without standards is why Gracia argues for a “strong proceduralism" that must take into account broad interests, including that of all of humanity[10].

Fifth, the puzzling relation of theory and practice led Toulmin a step further, noting that, not only does theory not matter, but *praxis* becomes paramount. In order to develop this point, I turn now to the post-modernist crisis in ethics.

CRISIS CONTEXT

The post-modernist movement creates a sense of crisis in bioethics and in ethics generally because of its questioning of any foundations for ethics.

In their book on casuistry, Jonsen and Toulmin alerted bioethicists to the thin support for principled ethics and held out a promise that casuistry and practical reason could provide a great deal of the consensus sought by public policy makers. The two ethicists based their argument for securing moral agreement among policy makers on the central role of *phronesis*[11]. This argument stemmed as much from the experience of developing principles for and positions about bioethical issues on the President's Commission for the Protection of Human Subjects of Biomedical and Behavioral Research that began meeting in 1974 and, in its early years, issued the Belmont Report. Both Jonsen and Toulmin served on this Commission.

Before writing with Jonsen about casuistry, Toulmin argued in his pathbreaking essay on how medicine saved the life of ethics that he was struck by the remarkable way the commissioners were able to agree on complex and delicate ethical issues. They did this, not by appealing to ethical principles, but through close attention to the texture of specific types of cases. In other words, persons of disparate backgrounds and convictions were able to weigh and balance the important values in different cases despite their different reasons for doing so. As Toulmin noted, "Only when the individual members of the commission went on to explain their own particular reasons for supporting the general consensus did they begin to go seriously different ways[12]. This primacy of practical reasoning over moral theory is central to my thesis as well[13].

Separately, Jonsen, Toulmin, and MacIntyre sought to develop a *phronesis*-based ethics centered on exercising "judgments in particular cases."[14] The very intransigence of the particular means that one cannot explain *phronesis* in procedural terms. Aristotle is quite explicit that the complexities of particular cases make them impossible to capture in any of the best-available rules. Thus, the very requirement to exercise judgment arises in the absence of objective certitude that even the mathematical sciences have difficulty providing[15].

My argument enters at this point. A search for foundations for clinical ethics must turn to the experience of patients, physicians, and other care givers in medical practices and in terms of their developing and evolving existential and experiential *a prioris*. Thus, there can be substantive goods discoverable through practical experience and wisdom which can form the basis of consensus or stronger recommendations. Before exploring the possible source of such goods, 1 will now turn to clarifying key ideas of postmodernism, antifoundationalism, and the problem of competing ethical theories in bioethics.

POST-MODERNISM

There are as many understandings of the movement we loosely call "post-modernism" as there are proponents for it and opponents to it[16]. In general post-modernism is a reaction to systems of thought that brought about the modern era, the industrial and now post-industrial age. Post-modernism is both a moral and a political reaction to authority, the authority of systems and those that act in their name, as well as the authority of rationality, standards and norms. Although there are many ways to characterize this movement, the principal convictions of post-modernism are that all philosophical systems are suspect. Moral and cultural pluralism must be recognized, and even appreciated, such that no overarching standards of conduct or "objective morality" are possible. There can be no foundational proposals for the theory of any human activity, much less a public policy[17]. All of these convictions, in turn, rest on a belief that the industrial ethic and the appeal to reason and natural law found in the Enlightenment project is dead[18].

Too, behind much of post-modern thought is either a nostalgia for a presumed, somewhat romanticized past[19], or a delight in skepticism that ignores its obvious violation of the principle of contradiction. Nostalgia possesses, in the words of Tamara Plakins Thornton, "bittersweet memories of a past when moral certainty still seemed possible"[20].

The skeptic revelers, by contrast, seek a clean sweep of outdated moral certitudes as a way of life and thought for coping with a pluralistic world. If it is a contradiction to speak apolitically by saying that nothing can be spoken about with certainty, then so be it.

A third group of persons who might accept the moniker of being "post-modern" are neither nostalgic Romantics, nor intensely skeptical theorists. They are simply cautious naturalists or pragmatists[21]. They are aware of the limitations of their own knowledge and beliefs, and keep open their thinking to new possibilities and discoveries [22].

Post-modernism arises not just in the recognition of multiculturalism, but in the egalitarianism of that recognition. In the United States, for example, until the 1920's, the majority considered its own, white and Nordic culture to be superior to other minority cultures. About that time a shift occurred such that various cultures were considered of equal value. One's preferences were held to arise without apology from feeling comfortable about one's own roots rather than from a judgment of superiority about one's cultural perceptions over another[23]. The problem engendered by this move is that one must accept relativism as a necessary adjunct to a multicultural sensibility (i.e., there can be no dominant or "primarily" valid viewpoint). One is then left without cross-cultural standards or any valid moral principles. Individual roots provide the only moral norms for a person's behavior. Anything becomes possible beyond these encouragements and restraints[24].

As noted, one characteristic underlying all of postmodernism is an intense suspicion of any rational effort to ground ethics in cross-cultural or trans-historical realities (Norris 1993). This is played out in bioethics in many ways. A good example right now is the effort to establish standards for bioethics consultations. What is the ethics consultant doing? If he or she is bringing to bear on a real tragic circumstance of human suffering a set of moral and objective standards in order to make a recommendation, then that is seen as "imposing" moral values on others, and is *de rigeur* out of touch with post-modern sensibilities. Yet if he or she does not bring some viewpoint to the negotiations, or has no aim in facilitating a good outcome, then why invite such a person to consult at all?[25] Is not bioethics supposed to help make our moral judgments internally coherent as well as rationally justifiable, and to provide a guide for sound public policy?

The focus on an awareness of moral and cultural pluralism also leads to profound questioning of fundamental assumptions of medicine itself[26]. Some of these assumptions have been held for centuries, e.g., the sanctity of human life, the goal of the preservation of life, and professional commitments to altruism. The post-modern critique requires more than a critical reflection on competing theories[27]. If we take the post-modern challenge seriously, re-iterating rational grounds for bioethics and the moral theory of medicine is insufficient. Something else is needed, and that something seems to reside within individuals, not theories.

COMPETING BIOETHICS THEORIES

When modern, secular bioethics began about 30 years ago[28], the hope was that by inviting the humanities into medicine and medical education, the humanistic aspects of medicine that had suffered so much under the success of scientific models of curing patients might be resuscitated and rejuvenated[29]. While that hope has been at least partially fulfilled, an unforeseen consequence has also occurred, one that is more detrimental than helpful to medicine. That unforeseen consequence is the importation into medicine of all the quandaries that possess modern philosophy, especially the competition among moral theories for ascendancy that we saw underscores the claims of post-modern thought.

In the last decade or so there has been a profusion of theories purporting to provide content, methodology and justification for biomedical ethics[30]. These competing theories are often contradictory and mutually exclusionary. As a result, students and practitioners alike are left with the impression that there are no "right answers" in bioethics, and that one may argue like the Sophists for the validity of any position by choosing what one wishes from the cafeteria of bioethical theories.

This is more than a matter of academic preference. What one believes to be right and good, and on what grounds, eventually shapes professional conduct and medical practice and, thus, the welfare of sick persons as well as the direction of public policy. Everyone has a stake in the outcome, and therefore, a stake in the convictions of practitioners. While every individual in society has moral convictions, all can agree at least that a goal of public policy would be to protect as many of these convictions as possible by fostering the liberty interests of all citizens, even though the public policy would then permit some violations of deeply-held convictions of some individuals within the society. Thus, the public policy principle of tolerance and respect for "liberty interests" would recognize moral pluralism while providing a bridge to the traditional goals of medicine. This may be one "way out" of the postmodern dilemma, and it turns one to the major point of the essay, the role of practical reason or practical wisdom in this effort, a turn away from abstract systems known by the "clever" and towards the daily moral judgments of everyday events.

PHRONESIS AS A WAY OUT

For some time now, ethicists have been arguing that behind ethics, at its foundation, is imply assertion and counterassertion. A good example of this claim is that made over 20 years ago by Alasdair MacIntyre[31]. At the time his claim arose from a consideration of pluralism. MacIntyre's more recent thinking has led him to consider spheres of moral enquiry in which certain values hold sway, but outside of which and across which they do not[32].

An even more radical rejection of morality comes from ethicists today who hold that bioethics is futile and misguided. Anne Maclean, for example, in her scathing attack on those who claim that ethical problems are in principle resolvable, argues against medical ethics as practiced by utilitarians. The purpose of utilitarian, indeed of all medical ethics, is seen as resolving problems raised by medical practice. Maclean says:

> The objection I wish to make to the bioethical enterprise is a fundamental one. it is that philosophy as such delivers no verdict upon moral issues; there is no unique set of moral principles which philosophy as such underwrites and no question, therefore, of using that set to uncover the answers which philosophy gives to moral questions. When bioethicists deliver a verdict upon the moral issues raised by medical practice, it is their own verdict they deliver and not the verdict of philosophy itself; it is their voice we hear and not the voice of reason or rationality.[33]

Here we stare directly in the face of post-modernist thinking. If this is to be considered a form of antifoundationalism, MacLean would argue, as she does in an oldfashioned Pyrrhist way, that at the foundation there is no truth of the matter in ethics. At best what we have are attitudes, not reasons, for why we do or should do certain things.[34] This view has much in common with the philosopher, Bruce N. Waller. Waller holds a version of "non-cognitivism" that claims that

> ... when ethical disagreements are run to ground in the search for resolution, ultimately there will remain only basic value preferences that cannot be rationally justified, and alternatives to which that can be favored without violating reason.[35]

"Basic value preferences" perhaps may not be able to be rationally justified, but they do stem from experience in a practice, and as such, emerge from a store of practical wisdom or *phronesis* of many individuals.

Aristotle's *Nichomachean Ethics* is grounded in his notion of *phronesis*. Indeed, this notion suffuses his psychology, political philosophy, and his notions of friendship and the common life. At the heart of *phronesis is* the notion of a role-model teacher, a person who possesses the wisdom required and can teach younger persons by doing rather than by words alone.

Today's bioethicists tend to recognize the importance of the idea, especially since it grounds ethics within a particular context, family network, social and cultural support system, yet parallels W.D. Ross' notion of "moral instinct" either as a rational judgment or rational faculty. Nonetheless there is a corresponding post-modern tendency to dismiss *phronesis* as elitist, since Aristotle argued that only the person with a complete moral life of virtue could exercise this virtue as well. Left over after such a dismissal of the *virtue* of prudence is the rational facility of practical reason alone.

The elitist dismissal is puzzling. Are not some positions better than others, some persons held in more esteem than others for their probity or courage or dedication? They are indeed, "better" than others in some respect. Aristotle's concept rightly implies that some moral judgments and some moral acts are better than others. In this sense, then, ethics itself implies an "elitism" of moral character. Nonetheless, in a properly functioning community, pre- or post-modern as it may be, there are many gifts and many to admire, one for humility, another for probity, another for courage, still another for wisdom. There is nothing undemocratic about this. Indeed, respect for individuals and their talents, their very individuality, moral and otherwise, is the heart of a democratic community.

Throughout all cultures and according the mores of each time and place, the most important virtue for human beings was moral wisdom. For the most part this was based on a shared conception of the Good, such that communities could train, support, promote, and honor their citizens without a great deal of self-critical analysis about whether the contents of the Good were justifiable.[36] Even in the advanced civilizations of Greece and Rome, with their explicit debates about the Good, unanimity existed that there was, indeed, such a reality, and that morality was achievable in pursuit of it.

Postmodernism directly challenges both convictions: first that there is an objective, publicly accessible Good, and second, that morality consists in pursuing such a Good. In its stead, the moral life is said to consist in devising one's own plan for a morally good life. In the absence of a generally accepted notion of the Good, pursuit of one's own plan is incredibly difficult to achieve. Confronting the obstacles requires a strongly motived, autonomous person, with great gifts, among them the thorough exercise of practical wisdom.[37] So by turning away from a publicly shared good to a private moral imagination and public courage, the notion of *phronesis is* hermeneutically reinterpreted to stand for individual moral probity, no matter what the goal, rather than a unique networking judgment, a habit of conjoining the abstract and socially shared conception of the Good with the everyday and practical exercise of it. As St. Thomas Aquinas argued, practical reason is "*quasi conclusio syllogismi practici*", like a conclusion of a practical syllogism.

Despite taking the high moral ground, the egoist position actually succumbs on at least two counts. First in the absence of a shared good the construction and pursuit of one's own plan is intensely solipsistic, such that all forms of reprehensible social behavior, up to and including the Holocaust, could be justified as just such an "authentic" pursuit of a plan. After all, as Erich Loewy has argued against "care ethics," the Nazis truly cared about their pograms.[38] Second, the requirements of moral imagination, courage, and dedication to one's private life plans and goals may produce saints, but it will also produce sinners -- those who singlemindedly pursue ethnic cleansing or kill for a Chicago White Sox jacket worn by a rival gang member or an innocent teen. Even this elemental observation betrays the need for objective standards. As I understand postmodernism, however, it is not the fact that we do have standards that puzzles them so much as the lack of rational justification for such standards and rules when subjected to the lense of honest multicultural appraisal.

Realizing this, most postmodernist ethicists require that decisions must be checked through peaceable dialogue and only those decisions reached by consensus might be judged to be appropriate ones.[39] At this point, however, one is quite rudely thrown back out onto the community, at least the community of dialoguers, and one's thinking is subjected to at least minimally communitarian checkpoints. The moral is necessarily linked to the political, as Aristotle held. Further assumptions about the Good reappear in such accounts as "side constraints," or conditions of possibility for the such moral dialogue, respect for pluralism, respect for persons, authentically listening even in disagreement, respect for liberty interests, and not violating the autonomy of another person unless he or she consents. These requirements for dialogue and/or a peaceful social existence, in a word, civility, are actually experientially constructed "goods" arising from activities in public life.
We can now examine the element of practice in *phronesis.*

PRACTICE

Praxis is an essential feature of *phronesis.* What is important about the virtuous or excellent individual is how that person functions day to day, gradually through experience developing the practical wisdom that others around him or her may lack to some degree. A virtuous person is not one filled with abstract knowledge about ethical theories, but one skilled at making prudential judgments for the good of others, oneself, and the community.

As St. Thomas Aquinas says in this regard,

> ...Prudence, which is right reason about things to be done, requires that a man be rightly disposed with regard to ends; and this depends on rightness of appetite.[40]

How does "disposition" and rightness of appetite come about?
It comes about through at least three means (a third means is also proposed by religious faiths, namely Divine inspiration, or assistance in the form of grace and the supernatural virtues).[41] For Aristotle, the first is the innate psychological structure of persons such that they have appetites or emotions that impel them towards particular goods or rewards. These appetites in our nature will control us without a more powerful direction from our higher intellect and will. The second means to acquire the virtue of prudence is through precept and example in the community. And a third one comes through reflection on one's choices in life and resolutions to improve these choices next time through repeated practice.

All three pathways work together in complex ways. Practical knowledge is experiential, progressive, and developmental. These characteristics, it seems to me, are lacking in the denuded form of *phronesis* seen, not as a virtue, but only as a reasoning facility. *Phronesis* comes about from doing right things and wrong things in life, and learning from them. Moral beliefs, too, arise from this source of practical knowledge, as do theories, principles, and rules (Wallace 1996). From this collected experience we are able to raise and train our children and provide for social stability. In medical ethics, inductive methods of analyzing particular conundrums lead to wisdom in clinical ethics and medical judgment.[42]

A point to be pressed here is that particular practices themselves, like medicine, acquire a collection of norms and standards. The justification of these norms and standards does not lie, as postmodernists rightly argue, in systems or theories of ethics, but rather in the practical realm of "doing" medicine, the doctoring and patienting activities that bring about healing. 1 will come back to this practical realm in my final section.

Glenn McGee explores this point by examining the problem of learning clinical ethics. In asking himself how one learns clinical ethics, he notes that moral learning takes place in any field through mentorship and practice. Indeed, as already noted, Aristotle's conception of *phronesis* requires this mentorship. The excellence of any person is tied to the long-term behavior of that person functioning in the community and society. When one acknowledges the moral excellence of any person, especially physicians, one attributes to them the wisdom of experience and practical judgments made every day, about mundane as well as the occasional dramatic matters.[43]

Dan Davis argues that while clinical reasoning and clinical medicine are sometimes referred to as a *techne* (or art) or as an *episteme* (or science), the best construal of clinical medicine and reasoning is as *phronesis.* While portions of the discipline of medicine can be described as an art or a science, the best paradigm is that of practical reasoning, since this notion fits better with the ways of knowing and doing carried out by the physician in a relationship with the patient. Davis'argument is anchored in our philosophy of medicine as a healing relationship conducted for a right and good decision for the patient, i.e., grounded in a beneficent professional healing encounter with the patient.[44]

IN SEARCH TO THE FOUNDATION: A PRIORI STRUCTURES AS A RESPONSE TO POST-MODERNISM

Most efforts at correcting anti-foundationalism compound the problem by not accepting the view that ail foundations are relativized; instead they have focused on rearticulating and redefending epistemological or metaphysical foundationalism.[45] Yet there are important arguments to be considered in ethical foundationalism as well.[46] A good example of an effort that skirts these thorny pathways is Ruth Macklin's book *Against Relativism.* In that embodied book she argues for a distinction between absolute truth (which does not exist) and universal truth (which, she proposes, does exist). This permits her to condemn certain medical practices that violate a universal truth in different ways in different cultures, without having to appeal to a system of absolute truth or to metaphysical or natural law principles.[47]

This struggle to find universals in the midst of diversity does not end the problem however. As Engelhardt notes:

> The foundational difficulty in establishing by discursive rational argument any content-rich account of justice, fairness, or public reason is that it presupposes agreement regarding background moral premises, rules of evidence, and rules of inference. This is precisely where agreement does not exist. As a consequence, the field of bioethics is characterized by deep, persistent, and often vehement disagreements...[48]

What steps can be taken regarding the impact of postmodernism on bioethics? First, the post-modernist critique of a Cartesian certitude in philosophy, ethics, or indeed, in any form of human knowledge should be taken as valid. It is helpful to face honestly the slippery foundation of all truth- or fact claims. This is especially true in ethics. Even Kant, the most rigorous of all moralists recognized this by holding that we do not really know *"Das Ding an Sich"* and that in ethics, the fundamental principle, the categorical imperative, requires us to "act as if." Kant once said, "Out of the crooked timber of humanity no straight thing was ever made."

The reason that there can be no Cartesian-like principles and norms in ethics is not that there can be no principles, norms, and standards at all, but rather because all reality, and our concepts of that reality, is other-referent. By that I mean that it is relational, having built into itself referents to other realities from which it came, to which it currently attends, and for which it will dissolve. In our conceptualization of these realities we cannot even imagine even abstract truths that do not relate to one another. In metaphysics, the good, the true, and the beautiful, are all transcendentals, i.e., they cannot be defined without reference to one another. In Einsteinian geometry and in relativity theory, the curvature of space means that no item can be defined without a referent to another. In mathematics, as Gödel's theorem attests, no purely deductive mathematics is possible, since for every “x" there must be a "y" interpreting the meaning of “x”.[49]

A more complete analysis of this phenomenon would lead to metaphysics of relationships. All reality is related to all other reality to such an extent that it cannot be defined without that relationship. To borrow a term from classical metaphysics, then, all reality is transcendental since it always refers beyond itself. This "arrow" of being implies that the existential structures of finitude, illness, death, and healing are also relationships, a point I develop in the penultimate section of this essay.

Turning to ethics we could not imagine any choice that does not involve a balancing of goods or cherished values. There is no "one good" in which all other goods are reconciled unless we turn to theology and belief. As Sir Isaiah Berlin said:

> The notion of the perfect whole, the ultimate solution, in which all good things coexist, seems to me to be not merely unattainable - that is a truism - but conceptually incoherent; I do not know what is meant by a harmony of this kind. Some of the Great Goods cannot live together. That is a conceptual truth. We are doomed to choose, and every choice may entail an irreparable loss.[50]

Hence, the rich conception of practical reason lies at the heart of ethical decision making. Not only the reasoning that is so essential to balancing the goods in this particular situation, but also emotional maturity and integrity is required to appreciate the gains, and to grieve for the losses during one's acceptance of the choice as it proceeds to affect the rest of one's life. That is why the meaning of *Phronesis* is essentially that of a virtue because it requires practice, and it signifies an habituation towards and development of one's standards and norms. Further, the choices made are incorporated into one's life and other judgments by which values and measured, weighed, and acted upon. Further still, the balancing of values occurs within the social context such that others are affected as well. The best post-modern ethics must therefore be one of connectedness, not just *difference*[51]. Second as a social judgment, practical reason is nothing less than the same virtue in lawmakers who have the power to enforce a judgment for the common good. As Aquinas argued, law itself is defined as a kind of public *phronesis,* a power of practical reasoning lodged in public authorities and/or legislators who can enforce standards on the community.[52] Aquinas shared with Aristotle this vision of law being a rational application of a universal rule to particular circumstances. He says:

> The rules of law and justice are in each case related to the actions performed in conformity with them as is universal to particular, for the actions are many, the law governing them only one, being universal.[53]

Traditionally the structure of social order has been seen to reflect some order in nature itself. Thus, the development of a natural law theory for social and personal order seems to rest on a psychological need in individuals to find some order in the social and natural chaos that surrounds them.[54] The ultimate interpretation of Aristotle's notion of rationality in law and ethics comes in Aquinas'view of a purposive universe in which all creatures are providentially governed.[55] This is precisely what is denied in post-modernism. Yet appeal to rationality still grounds most of our legal theory. In international bioethics, for example, the International Bioethics Committee of UNESCO bases its arguments about the right not to be discriminated against on the basis of genetics on universal principles like "the intellectual and moral solidarity of mankind" and "the common welfare of mankind" because these, in turn, are "indispensable to the dignity of men and constitute a sacred duty which all the nations must fulfill".[56] Rationality is essential for the inductive process itself since there is a fit between innate psychological capacities and the causal structure of the world[57].

A third point is relevant here. The point made by post-modernists that the Enlightenment Project is dead is an important one. By this not only meant that there will be no chance to develop a completely rationally coherent basis for ethics and law in society, but also that such efforts produce deadly results. Ironically, the Enlightenment was itself a reaction against Absolutism, most particularly politically absolutist claims made by Kings in the mid-1600's. Grounding authority in individuals, their rights, and their inherently rational nature, was a radical break with the notion of external authority in society[58]. But the Enlightenment only transferred authority; it did not eliminate it entirely. Individuals appealed to the natural order, to "Nature and Nature's God". The deadly results are formed by the systems of thought in which reason dominated all other versions of reality, eliminating the emotional, irrational, impulsive, and incoherent features of human beings and nature itself.

An example would be the habit arising from the Western manner of solving problems by objectification, breaking them into smaller, more manageable parts, and then manipulating and commodifying them[59]. This general description fits our search for the genetic basis of disease, for example, transplant technology, or the approach made by a clinician towards solving an appendicitis attack (without the commodification step, of course). While this approach works out for the good in many cases, the habit itself has led to horrible consequences more specific to bioethics, say with respect to prolonging a dying patient's life, or in reproductive technologies where living matter is treated as an object of study[60].

Behind this particular scenario is a more general danger. Accepting any one vision means to accept what passes for objective verity in its pursuit of The Good. The post-modern warning is that this truth and that good are only based on subjective preferences or unexamined cultural assumptions (in the final analysis). Not to recognize this contingency at the root of all systems is to suspend doubt and barrel down the road of progress provided by the Enlightenment pursuit of objective reality. What practical reason supplies to this broadly-painted dialectic is a discovery of truth and goods within the concourse of human relationships and experience. Neither totally objective nor totally subjective, the ethics that arises from *phronesis* develops its wisdom over the course of a life, for individuals, and the course of a society and civilization for a community.

But what experience is present that might transcend individual preferences with regard to bioethics? In medicine, and thereby in medical ethics, there is the existential principle of human finitude (Pellegrino and I are working on developing this principle as an "experiential *a priori*" in medicine and bioethics as one response to the post-modernist crisis). We all share in this finitude and its effects, illness, decay, health itself (which is always a temporary condition), disability, vulnerability, and death. These existential conditions are not alien to the post-modern speaker and listener, but constitute the conditions of physical existence that define our lives. While we can debate the proper responses of a community to these conditions, they are undeniable facts of life that medicine addresses. The responses to these conditions create the standards to which we must and should appeal as fellow humans in any ethical discussions about medical theory and practice. These standards arise from experience with making practical judgments about values in that medical context.

These are, then, the irreducible medical realities with which every bioethics theory must deal. They are common to the human experiences and predicaments that have given birth to the need for medicine in every time and culture. These realities are the phenomena of medicine, those that make it a special kind of human activity, with special and distinctive moral characteristics[61]. These standards are more than just role-specific duties taken on through public profession and promise keeping[62]. They are rooted in vast and universal human experience.

Given these reflections about one path out of the post-modernist challenge to truth in clinical ethics, a path directed by Aristotle's *phronesis,* what contributions can it make to preserving some standards in ethics? Instead of looking to either defend or reject the "correspondence theory" of truth in ethics, one can reinterpret that theory such the truth in ethics corresponds to learning and to developing in accordance with one's being and relationships. A search in this direction is not an abandonment of the search for truth, but rather its proper path. Further, as Bernard Williams argues, some concepts become "thicker" than others to the degree to which others converge with it as it moves towards objectivity[63]. The thicker the concept the more validity it has for general application in the sense that it has been tested and converges with other experiences. These experiences are still time- and culture-bound, but as they pass from century to century they acquire a validity that approaches "objective" truth. In my view, this is the greatest and only form of "objectivity" that ethics can approach: "Generally for the most part true."

Built into this level of abstraction is hard-fought contingency flagged for us by post-modern thinkers. No principle in ethics should be posited without its individual and cultural referent. An example might be the rule against killing. The rule is a short-hand for a statement that should read, if spelled out, "Generally for the most part it is evil to kill. This rule arises from our personal and cultural experience that killing destroys both the sacredness of the life of the victim and the boundaries that each individual should place on his or her own being in consort with others in the community." I call this the "experiential a priori" that should be present on the table of all negotiation and discourse about values.

This inductive process leads to a practical truth, then, a truth of standards inductively drawn that are for a single individual "generally for the most part true" based on her life-experience, and her relations among family, friends, acquaintances, business, social, and cultural worlds. The broader her experiences and more complex her choices, the "wiser" she becomes compared to others with less practice. Similarly, a society can grow (and decline) in practical wisdom for many of the same reasons.

Turning these points to bioethics and postmodernism, there are several problems with non-cognitivist views of bioethics I took up earlier in the discussion of post-modern challenges. The first is McLean's identification of bioethics with utilitarianism. Only a few bioethicists can be called such, Harris, Rachels, Singer, and R. M. Hare, for example. One could object to MacLean's arguments on their behalf, as did John Harris[64]. Harris holds that her view is fundamentally flawed. One might expect this critique from him, as he is one of the principal thinkers against whom McLean argues.

Without becoming embroiled in the arguments, this debate is interesting for my purpose because it sheds light on an almost mechanistic process of cataloguing and resolving ethical questions in medicine. In their efforts to appeal to reason, utilitarians are vulnerable to the critique that the appeal to reason is really an appeal to schema they have designed that precisely take away the moral struggle. Consider, for example, Hare's proposal of a canon of moral rules people must follow to think rationally about ethics. He claims that everyone who follows this method correctly, MacLean argues, "will come to the same moral conclusions."[65] One can readily see why MacLean objects to this kind of rote reasoning. As Cornford put it in his spoof of academe, "Plainly, the more rules you can invent, the less need there will be to waste time over fruitless puzzling about right and wrong"[66]. Of course we would like standards gleaned from past experience, as conscience provides for individuals, but these standards should not be grounded so much in the more abstract rules of reasoning well about ethics as in prudential experience. This is the realm of prudence.

Second, MacLean concludes, in the words of another reviewer, Elizabeth Telfer, that 'Moral philosophy's proper role in medicine is only one of clarification"[67]. That is to say, philosophers should analyze problems in everyday language, and then propose to moral agents, such as patients, or policy makers, the many possible answers. The moral agents would then make up their own minds.

Thirdly, the "value preferences" that MacLean finds at the foundation of bioethics, or that noncognitive ethicists accept as present at the root of moral debates, are not simple personal, and subjective assertions if they are grounded in the values of patients and doctors about healing illness and disease, the existential a priori of human finitude I continue to emphasize. Such "preferences" are cognitive and experiential a prioris for a moral philosophy of medicine. They stem from centuries of medical practice, centuries of physicians and patients caring for illnesses common to all persons, across time and across cultures.

Further, the utilitarians are correct that there are rules of right thinking. If there are no foundations, for example, some irrefutable principles that are necessary or necessarily true, such as Aristotle's principle of contradiction, or the law of the excluded middle term, then reasoning itself becomes chaotic. There must be a necessary set of principles that ground all claims to know or to ascribe meaning. Antifoundationalism contends about knowledge and reasoning that there is no set of principles inherent in knowing, and that all such have been derived from the human act of achieving knowledge itself. This point is valid, as I have argued, since these principles have evolved from the habits of thinking and reasoning over time and have been tested as valid by human experience. Antifoundationalism in this category is important for a moral philosophy of medicine because it rests on rejection of the one-dimensionality and reductionism to universally abstract principles of thought and action.

STRUCTURES OF HEALING AS FOUNDATION OF CLINICAL BIOETHICS

The argument thus far leads to the conclusion that the moral foundations of health care practice are less to be found in respecting some objective standard ethical theory, or balance of rules or *prima facie* principles, or even in an account of autonomy such as free choice, and more to be found in the ecology of the person, society, and natural world[68]. These are the non-philosophical indicators of autonomy as a richer feature of all human relationships, not to speak of foundational relationships, of which the sickhealer relation is just one. Let me explain this further.

To respond to postmodernism and to reflect on the practice of clinical ethics leads us to the foundational realm of human existence that Wartofsky called the "rich life." Without this deeper analysis, one is condemned to consider ethical theories only based on narrative, caring, or other more everyday ethical theories, or on the more objective, scientific level of abstractions. The latter is abhorred by the postmodernists and the former by those for whom there must be an objective morality to check the violent tendencies of human life and culture today. There is another level of experience, however, that could ground both the everyday *Lebenswelt* and the objective, scientific world. I call this the realm of existential structure[69]. Bergsma and I call this the ecological model of the moral foundations of healing that transcends exchanges and resistances to power. Noting how an academic understanding of autonomy and insufficient clinical experience can torpedo the healing relationship, we argued that clinical experience with different coping styles of patients is a window on the deeper existential structural realm that the doctor and patient must share.[70] [71]

CONCLUSION

A mistake of many arguments against postmodernism is not to grant its assertion that something radical has happened in recent human history. In turn, the postmodernists perhaps misidentify the source of that rupture with the past. By arguing that pluralism, multiculturalism, and the particularity of moral discourse collapses the objective, rationalistic ethics of the Enlightenment, postmodernists too easily dismiss the need for some foundation for any moral discourse, even while they engage in it.

This essay has suggested that there is a ground for all disciplines in what I have called foundational relationships. Further I argued that the foundational relationship for health care is the sick-healer relationship. The fact that this foundational relationship takes different forms in different cultures does not mean that it does not exist at all. The richness of this structural ground was not explored in this essay, except to suggest that it may include empathy, co-suffering, transfer of the power to heal to the healer[72], and the reintegration of the self and its world[73]. These are moral acts because of the good embodied in the relationship. From the moral acts can be developed axioms that can function as norms for the healing relationship Bergsma and I call the ecological model of the doctor-patient relationship. Note, too, that Pellegrino and I developed three such axioms from the goal of medicine to heal, in our earliest joint work[74].

Much more needs to be done to spell out all the conditions of possibility, the formal causes in Aristotle's terminology, of the moral good of healing, and the foundations of clinical bioethics.

David Thomasma †, Professor and fr. English Chair of Medical Ethics, Neiswanger Institute for Bioethics and Health Policy, Loyola University of Chicago, Medical Center, U.S.A..

NOTES

[1] Michael L.(1998), *Blue sky like water*, In Anon, ed. *Shorts.* London: Granta Books, p. 224.

[2] Levi P. (1996), *Survival in Auschwitz: The Nazi Assault on Humanity,* New York: Simon & Schuster.

[3] Russell B. (1969), *The ABC of Relativity,* London: G. Allen & Unwin.

[4] This is why the new journal published by the Fondazione Lanza, *Etica per le Professioni,* is *so* important today. Issues such as privacy, to which the first issue was devoted, cut across disciplinary, social, and national boundaries.

[5] Thomasma DC. Bioethics and international human rights. *Journal of Law, Medicine & Ethics* 1997; 25: 295-306. Also Baker R. A theory of internationalbioethics: the negotiable and the non-negotiable. *Kennedy Institute of Ethics Journal* 1998; 8 (3): 233-273; Veatch R. *Cross Cultural Perspectives in Medical Ethics.* Boston: Jones and Bartlett, 1989; Gillon R. *Principles of Health Care Ethics.* New York: Wiley, 1994.

[6] Morreim, EH (Ed) (1994), Bioethics, expertise, and the courts. *The Journal of Medicine and Philosophy*; 22 (4), pp. 403.

[7] Viafora C. (1999), Toward a methodology for the ethical analysis of clinical practice. *Medicine, Health Care and Philosophy*; 2 (3), pp. 283-297.

[8] Simon H. (1957), Models of Man: Social and Rational,- Mathematical Essays on Rational Human Behavior in Society Setting, New York: Wiley.

[9] Viafora C. (1999), p.285.

[10] Gracia D.(1989), *Fundamentos de bioetica,* Madrid: Eudema.

[11] Jonsen A, Toulmin S. (1988), *The Abuse of Casuistry: A History of Moral Reasoning,* Berkeley, CA: University of California Press.

[12] Toulmin S. (1982), *How medicine saved the life of ethics. Perspectives in Biology and Medicine*, 25, pp. 736-750

[13] However, Beresford points out that in order for *phronesis to* secure the agreements promised by.Jonsen and Toulmin, and needed in health care ethics, it must pay attention, not only to details of cases, the concreteness stressed throughout this essay, but also must focus on the central substantive goods in the case. It is the latter, he argues, that continue to be elusive; Beresford EB. Can phronesis save the life of medical ethics? *Theoretical Medicine* 1996; 17 (3): 209-224.

[14] MacIntyre A. *After Virtue: A Study in Moral Theory.* Notre Dame, IN: Univerity of Notre Dame Press, 19 8 1; *Whose Justice, Whic Rationality?* Notre Dame, in: University of Notre Dame Press, 1988, p. 117.

[15] Aristotle (1923), *Nichomachean Ethics,* London, Loeb Classical Library, p. 109 4b.

[16] Hoy DC, McCarthy T. (1994), *Critical Theory,* Oxford/Cambridge, MA: Blackwell.

[17] Critchley S. (1992), *The Ethics of Deconstruction: Derrida & Levinas.* Oxford/Cambridge, M: Blackwell.

[18] Beiser FC. (1997) *The Sovereignty of Reason: The Defense of Rationality in the Early English Enlightenment*, Princeton, NJ: Princeton University Press.

[19] Consult, for example, the first endnote in Engelhardt 1991, where he bemoans the losses of certainty and ritual in the Roman Catholic tradition. There is a distinct feeling in reading his work that Engelhardt wishes secular pluralism were not the case, but is ruthlessly honest that it is, and that his thinking takes the loss of the Enlightenment project seriously; see also, Engelhardt HT. The crisis of virtue: arming for the cultural wrs and Pellegrino at the limes. *Theoretical Medicine* 1997; 18, pp.165-172.

[20] Thornton TP. *Handwriting in America: A Cultural History.* New Haven, CT: Yale University Press, 1996; Mallon T. Minding your "p's and q's." *The New Yorker* 1997; 72 (45): 80.

[21] Nielsen K. (1997), *Naturalism without Foundations,* Amherst, NY: Promethesus Books.

[22] Horgan T, Tienson J. (1996), *Connectionism and the Philosophy of Psychology,* Cambridge, MA: MIT Press.

[23] Michaels MT (1995), *Our American Nativism, Modernism, and Pluralism.* Durham: Duke University Press.

[24] de Wachter MAM. (1997), The European Convention on bioethics. *Hastings Center Report* 27 (1), pp.13-28.

[25] See articles on ethics consultations in *Journal of Clinical Ethics* 1996, in particular Fletcher JC, Siegler M. What are the goals of ethics consultation? *Journal of Clinical Ethics* 1996; 7 (2): 122-126; see also Aulisio MP, Arnold RM, Youngner SJ. Can there be educational and training standrds for those conducting health care ethics consultation? In Monagle, JF, Thomasma DC (eds), *Health CareEthics: Critical Issues for the 21 Century.* Gaithersburg, MD: Aspen Publications, 1998.

[26] See my essay relating anti-foundationalism and any future philosophy of medicine: Thomasma DC. Medical ethical theories. *Textbook of Military Medicine.* Washington, DC: Office of the Surgeon General, forthcoming.

[27] Veatch RM. (1989), *Hospital pharmacy: what is ethics? American Journal of Medicine and Philosophy*; 46 (1), pp.109-115.

[28] This date is disputed by some - for a particular history of modem bioethics, see Jonsen A. *The Birth of Bioethics.* New York: Oxford University Press, 1998; also see Pellegrino ED. The metamorphosis of medical ethics: a 30-year retrospective. *Journal of the American Medical Association* 1993; 269: 1158-62.

[29] Pellegrino ED, McIlhenny T (Eds), (1982), *Teaching Ethics, the Humanities and Human Values in Medical Schools: A Ten-Year Overview,* Washington, DC: Institute of Human Values in Medicine/Society for Health and Human Values, p. 101.

[30] Thomasma, forthcoming.

[31] MacIntyre A. (1979), Why is the search for the foundations of ethics so frustrating? *Hastings Center Report* 9 (4), pp. 16-22.

[32] MacIntyre A.(1990), *Three Rival Versions of Moral Enquiry,* Notre Dame, IN: University of Notre Dame Press.

[33] MacLean, A. (1993), *The Elimination of Morality: Reflections on Utilitarianism and Bioethics,* New York: Routledge, p.5.

[34] Ibid: p. 35.

[35] Waller B. (1994), *Noncognitive moral realism. Philosophia*; 24 (1-2), p. 59.

[36] Charvet J. (1995), *The Idea of an Ethical Community,* Ithaca, NY: Cornell University Press.

[37] Kekes J. (1997), *Moral Wisdom and Good Lives.* Ithaca, NY: Cornell University Press.

[38] Loewy 199

[39] Engelhardt HT. (1996), *The Foundations of Bioethics,* 2nd Edition. New York: Oxford University Press.

[40] Aquinas T. *Summa Theologiae.* II, Iiae, q.57, a. 4.

[41] With respect to medicine see Pellegrino ED, Tomasma DC. *The Christian Virtues in Medical Practice.* Washington, DC: Georgetown University Press, 1996.

[42] Graber GC, Thomasma DC. (1989), *Theory and Practice in Medical Ethics.* New York: Crossroads.

[43] McGee G. (1996*), Phronesis in Clinical Ethics. Theoretical Medicine* 17 (4), pp. 317-328.

[44] Pellegrino ED, Thomasma DC. *A Philosophical Basis of Medical Practice.* New York: Oxford University Press, 198 1; and *For the Patient's Good: The Restoration of Beneficence in Health Care.* New York: Oxford University Press, 1988.

[45] Chisholm RM. (1982), *The Foundations of Knowing,* Minneapolis, NIN: University of Minnesota Press; Habermas J. (1988),*Nachmetaphysisches Denken: Philosophische Aufsaize.* Frankfurt: Suhrkamp.

[46] Tollefsen C. (1995), *Foundationalism Defended: Essays on Epistemology, Ethics and Aesthetics.* Bethesda, MD: Cambridge University Press.

[47] Macklin R. (1999), *Against Relativism: Cultural Diversity and the Search for Ethical Universals in Medicine*. New York: Oxford University Press.

[48] Engelhardt HT Jr. (1999), *Bioethics in the third milennium: some critical anticipations. Kennedy Institute of Ethics Journal*; 9 (3), pp. 225-244. Quote: 236.

[49] Wang H. A (1996), *Logical Journey from Godel to Philosophy,* Cambridge, MA: MIT Press.

[50] Berlin I. (1991*), The pursuit of the ideal. The Crooked Timber of Humanity: Chapters in the History of Ideas,* New York: Alfred A. Knopf, pp.1-19.

[51] Bauman Z.(1993), *Postmodern Ethics,* Oxford/Cambridge, MA: Blackwell.

[52] Aquinas II, II, Q. 9 1.

[53] Aristotle, (1953).

[54] Smith JC, Weisstub DN. (1979), *The evolution of western legal consciousness. International Journal of Law and Psychiatry* 2, pp.215-234.

[55] Nielson, (1997).

[56] UNESCO (1988), Universal Declaration on the human genome and human rights. *Journal of Medicine and Philosophy* 1998; 23 (3): 334-341.

[57] Kornblith H. *Inductive Inference and Its Natural Ground: An Essay in Naturalistic Epistemology*. Cambridge, MA: MIT Press, 1993.

[58] Academy of Humanism. *Challenger to the enlightenment: In defense of Reason and Science*. Amherst, NY: Prometheus Books, 1996.

59 James C. Sex and Reason: Bertrand Russel, passionate about both. The New Yorker 1996; 72 (38): 104-117.

[60] Crewdson J. US fires research over use of embryonic DNA. The Chicago Tribune Jan. 9, 1997; Sec. 1:4.

[61] Pellegrino, Thomasma, 1981.

[62] Thomasma DC. Promisekeeping: An institutional ethos for healtcare policy. *Frontiers of Healt Services Management* 1996; 13 (2): 127-143.

[63] Williams B. The truth in ethics. In Hooker B, *Truth in Ethics*. Cambridge: Blackwell Publishing, 1996.

[64] Harris J. The elimination of morality. *Journal of Medical Ethics* 1995; 21: 200-224.

[65] MacLean, 1993, p. 114.

[66] Cornford FM. *Microcosmographia Academia: Being a Guide for the Young Academic Politician*. 5th Edition. Cambridge: Bowes & Bowes, 1953, p. 10.

[67] Telfer E. The elimination of morality. *Philosophical Books* 1995; 36 (3): 204-206.

[68] Bergsma J, Thomasma DC. *Autonomy and Clinical Medicine: Renewing the Health Professional Relation with the Patient.* The Netherlands: Kluwer Academic Publishers, forthcoming.

[69] This tripartite division is based on Husserl's detailing of the *Lebenswelt* under abstract concepts, and Merleau-Ponty's later argument for a pre-conceptual realm he called "the-body-subject-in-the-world". The hyphens are used to indicate that this realm is an act, a perception, underlying subsequent actions and articulations of it.

[70] Katz J. *The Silent World of Doctor and Patient.* New York: Free Press, 1984.

[71] Bergsma, Thomasma, op. cit.

[72] Brody H. *The Healer's Power.* New Haven: Yale University Press, 1992.

[73] Bergsma J, Thomasma DC. *Health Care: Its Psychosocial Dimension*, Tr. By D. Thomasma, being a translation from the Dutch and expansion of J. Bergsma, *Somatopsychologie: op zoek naar psychosociale dimensies van het geneeskunde* (Lochem: De Tijdstrom, 1977). (Pittsburg: Duquesne University Press, 1982).

[74] Pellegrino ED, Thomasma DC. *A philosophical Basis of Medical Practice*. New York: Oxford University Press, 1981.

EDMUND PELLEGRINO

THE "TELOS" OF MEDICINE AND THE GOOD OF THE PATIENT

INTRODUCTION

"Do good and avoid evil". This is the ancient dictum of *synderesis,* and, implicitly or explicitly, the indispensable transcendental ground for any system of ethics of moral philosophy. This is because the good is the end or *telos* of moral science, that which gives it its distinctive identity among human activities. One may locate the good in many places - in natural law (Aristotle and Aquinas), in the will (Kant), in the affect (Hume), or intuition (Moore). But no ethics can avoid the concept of the good, since, without the good as *telos,* the word "ought" is without direction, and morality, itself, dissolves in the acid of skepticism.

Medical ethics has its own construal of *synderesis* in the first moral precept of the Hippocratic Oath. Here, it states:

> I will follow that system or regimen which, according to my ability and judgment, I consider for the benefit of my patient and abstain from whatever is deleterious and mischievous.... (Francis Adams's translation)

All the moral precepts of the Oath, the other deontological books of the Corpus, and the entire subsequent history of medical ethics are elaborations of this first principle. This medical *principium primum* enjoins upon doctors the primary duty of acting to benefit the patient, in a word, always to act for the patient's good. The patient's good is, therefore, the end of medicine, that to which medicine, by its nature, tends and that which gives it its definition.

Until the beginning of the modem era of bioethics, some thirty-five years ago, this notion of the primacy of the sick person as the good of medicine, the end toward which it should tend, would have been unchallenged. Today, this assertion is increasingly under attack by changes in moral philosophy as well as in our social and political mores. The ends of medicine, the good it should tend to, and, consequently, the ethics that should guide it, have become problematic. There is now a great danger that the good of the patient as the architectonic end of clinical medicine will be severely compromised or lost entirely.

Redefining the end of medicine has, thus, become the central problem in the philosophy of medicine. What for centuries seemed so certain is now problematic. Historians, sociologists, ethicists, physicians, policy-makers, and the public are asking the same questions: What should physicians be - primarily healers or servants of societal good, businesspeople, entrepreneurs, bureaucrats, scientists, *etc.*? How should medical knowledge be used, who should decide, how, and by what criteria?

C. Viafora (Ed.), Clinical Bioethics. A Search for the Foundations, 21-32.

These questions imply an altered conception of the good that medicine should serve, the roles physicians should play in serving those ends, and the ethical obligations that should bind them. Recently, these questions were examined seriously in a cooperative, international, interdisciplinary report under the aegis of the Hastings Center (Hanson and Callahan 1999). No consensus was reached on the goals or purposes of medicine. There was agreement, however, that the preferred method for determining goals is through some form of social construction, *i.e.,* by public dialogue, negotiation, or debate. With one exception, the contributors rejected the classical notion of ends in favor of goals or purposes. (Pellegrino in Hanson and Callahan 1999)

In this essay, I will attempt to recover the ancient notion of a teleological ethics. I will do so by turning my attention to ends, rather than goals or purposes, and by addressing four questions: (1) What are the roots of the current conceptual confusion and the need for redefinition of ends? (2) Within the framework of an Aristotelian-Thomist teleological ethic, how do we define the ends of clinical medicine? (3) How do we define the good as the telos of medicine? (4) What implications does the teleological approach have for the ends of the other learned professions - ministry, law, and teaching? (5) What are its implications for virtues and principles in medical ethics?

The conclusions I will reach are these: (1) The roots of the current confusion about the ends of medicine and the need to redefine them lie in the erosion of the classical conception of ends. (2) Recovering that classical conception leads to the conclusion that the end of clinical medicine is the good of the patient. (3) The good of the patient is a quadripartite idea, and this idea grounds the ethics of medicine. (4) The teleological approach can be fruitful in defining the ends of other branches of medicine as well as other professions.

SOME PROPEDEUTIC DISTINCTIONS

I will confine this inquiry to clinical medicine rather than "medicine" generically. By *clinical medicine,* I mean the use of medical knowledge and skill for the healing of sick persons, here and now, in the individual physician-patient encounter. Clinical medicine so defined is the activity that defines physicians *qua* physicians - and other clinicians - and sets them apart from other persons who may have medical knowledge but do not use it specifically in clinical encounters. Clinical medicine is the physician's *locus ethicus* whose end is a right and good healing action and decision. Moreover, clinical medicine is the instrument through which public policies come to affect the lives of sick persons. Finally, no matter how broad or socially oriented we make medicine, illness will remain a universal human experience, and its impact on individual human persons is the reason why medicine and physicians exist in the first place. (Cf. Hippocrates, On the Art)

Using clinical medicine as a paradigm case is not to neglect the other branches of medicine, each of which has its own distinctive end. Thus, for basic scientists, the end is the acquisition of fundamental biological knowledge of health and illness. This knowledge becomes a part of clinical medicine specifically when it is applied to the needs of a particular human being here and now. Similarly, preventive medicine has as its defining end the cultivation of health and avoidance of illness. Social medicine has its end in the health of the community or the whole body politic. When the knowledge and skills of any of the other branches of medicine are used in the healing of a particular person, then the ends of that branch fuse with the ends of clinical medicine. But in clinical medicine, the good of the patient is the end, *primus inter pares.*

Some of the difficulty of the Hastings Center group in arriving at a consensus arose because these distinctions were not made clearly enough. The group tended to expand the definition of medicine so broadly as to absorb or "medicalize" almost all aspects of life. Such an expansion defeats any attempt to define ends. It places ends in conflict with each other and weakens any attempt to establish a hierarchy of goods among the many ends "medicine" may serve. (Ingemar Nordin).

Finally, it is necessary to clarify my use of the term "teleological ethics." By this, I do not mean any form of consequentialism or its major expression in utilitarianism. Nor do I mean a simplistic biological teleologism. Rather, I refer to an ethic based in the notion of the good as the end of moral acts wherein "good" is defined in terms of the nature of the activity in question, that for which the activity exists. Such an ethic is the antithesis of an ethic of social construction in which the good is defined externally to the activity in question by what we wish or intend the activity to achieve. Elsewhere, I have spelled out the problems with a socially constructed notion of the goals, purposes, or good of medicine. (Pellegrino in Hanson and Callahan, 1999).

THE NEED FOR A REDEFINITION OF ENDS

Today's confusion about the ends of medicine and the need for their redefinition lies in the erosion of the Classical-Medieval notion of ends, their relation to the good, and the relation between the idea of the good and ethics. If the end of medicine is to be redefined, the ancient concept of ends must first be retrieved from exile by modern and contemporary philosophy.

As is well known, Aristotle begins the Nicomachean Ethics with the proposition that the good is that which all men desire. The good is the end or telos of human activity, and the end is that for which a thing exists, that which an act is designed to bring about. Ends are in the nature of things themselves. They answer the question, "What for?" (Jonas 1984, p. 52). We do not impute ends to things; they are not good because we desire them. We desire them because they are good. We may put things, like medicine, to certain goals and purposes, but whether these are good or bad uses depends upon whether they fulfill the ends for which medicine exists and which define it *qua* medicine. Aristotle uses the familiar example of medicine to illustrate his meaning of ends. Medicine, he asserts, is a *techné* whose end is health, just as the destination of a ship is the end of navigation. These ends are also the good for each activity since they are what men seek by engaging in the activity, and the activity is so structured that, by its very nature, it is ordained to the end that defines it.

Aristotle and Aquinas, in a similar fashion, were concerned chiefly with the larger conception of the good for humans as the end of human activity. Both structured their moral philosophies on the good as the end of human life. That end in its ultimate sense was, for Aristotle, a life consistent with the natural virtues, which led to happiness. For Aquinas, it was a life lived in accord with the natural and spiritual virtues which led to the beatific vision and fulfilment of the spiritual nature of humans.

Both Aristotle and Aquinas anchored the virtues in the ends of human life and the good. Thus, they linked metaphysics with ethics. The virtues became habitual dispositions to act in such a fashion that the end of human life, the good, that to which it tended by its very nature, could be attained. Both Aristotle and Aquinas used medicine as an example of a human activity with a definable end and good, a lesser good, of course, than the ultimate good of human beings as such. They defined the end of medicine as health, that toward which the activity of medicine tended, that which made it what it was and which distinguished it from other human activities.

Thus, in determining the ends and good of human life, and in the realm of lesser good in everyday life, ends and the good are intimately related. For our purposes in this essay, we can pass over such questions as: the apparent contradiction in Aristotle between the primacy of intellect or will in perception of the good; whether the good is determined because we choose it, or we choose it because it is good (cf. Plato, "Euthyphro," p. 391); and whether *telos* and end are coextensive in Aristotle (Hardie 1980, pp. 254-257). Those are important questions, but their answers are still within the teleological tradition.

However, from the late Thirteenth and Fourteenth Centuries to our times, the foundations for a teleological ethic have been seriously eroded. The Nominalists began the process by rejecting any grounding for universals or essence in the nature of things, thus disarticulating the connections between ends and the good. This process accelerated in the Eighteenth Century and has continued to the present. (MacIntyre 1966). The history of the erosion of the teleological ethic is too complex for citation here. Four conceptual shifts over the centuries seem particularly relevant: (1) The rupture between ethics and metaphysics in Kant's insistence on locating the good in the will and ethics in reason alone. (2) Hume's denial of any logical connection between "is" and "ought," between fact and value, and his preference for affect over reason in ethics. (3) G. E. Moore's declaration that the good is an indefinable quality. And (4) the discrediting of any stable foundation for moral philosophy as well as extreme skepticism about the possibility that moral truth is ascertainable by the use of human reason.

The upshot of all of this has been profound for moral philosophy, particularly for Anglo-American ethical theory. Emphasis on the good has been shifted to an emphasis on the right or to the act of commending or valuing out choices and preferences. Morality, itself, becomes the creation of our choices and the mores of a liberal society. The good is revealed in the choices we make

On the modernist view, especially in Anglo-American philosophy, moral choices are simply our preferences among the sentiments dominant in our society at a particular time. Our differences are differences of language or interpretation of a set of facts. This is the "foundation" for the now-dominant social construction theory of morals. It is a direct antithesis of the natural law tradition which recognizes a transcendent moral reality. The moral person is expected to discover that reality by the use of reason and to be guided by it. And, *pace* Hume, moral truth can be derived from existential fact.

THE END OF CLINICAL MEDICINE

Aristotle and Aquinas both define the end of medicine generically as health. Kass defines this end more specifically as the well-functioning of the human body. These definitions are general enough to encompass clinical medicine as well as the basic sciences related to medicine, including social and preventive medicine. But they need to be refined further for our project which is to define the ends of clinical medicine.

Clinical medicine, as we have defined it, centers on the clinical encounter between someone with medical knowledge and someone who needs that knowledge to restore health, the well functioning of the human organism which has been disturbed by dis-ease. For the sake of that end, a more proximate end must first be gained, that is, a right and good healing decision must be made and safely implemented by the clinician and the patient. This is the more immediate end of the clinical encounter. It is that which the ill or sick person needs and that which the clinician *qua* clinician is ordained to provide.

Medicine is, indeed, a *tekné* as Plato and Aristotle said. It is, in medieval terms, a *recta ratio factibilium,* a right way of making or doing a thing. Clinical decisions must, therefore, be technically right, *i.e.,* scientifically correct, but also morally good. This fusion of the technical and the moral good is effected in the interest of a vulnerable, anxious, dependent, exploitable person. This is the indispensable starting point of the further end of healing and health.

Healing is the act of acts whereby the disruptions of bodily, mental, and spiritual harmony of present health is disrupted by disease. If the patient is to be healed, the disruption of disease must be ameliorated and the violence lessened or, if they are irreversible, healing may still take place, although its focus will shift from curing to caring, coping, and alleviating pain and suffering. These are wounds to the patient's humanity which can be healed even when the patient is dying. Cure may become futile, but care is never futile.

The end of healing is the good of the whole person - physical, emotional, human, and spiritual. The good is, therefore, quadripartite, healing at four levels of human being and living: the lowest level is the medical good; above that is personal good as perceived by the patient; next is the good of the patients as a human being; and the highest good is the patient's spiritual good. The good of the patient is therefore an hierarchical good in which each level must be in proper relationship with the others.

DEFINITION OF THE PATIENT'S GOOD
THE MEDICAL GOOD

The medical good is that which relates most directly to the art of medicine, that part which is uniquely medical. The medical good aims at the return of physiological function of mind and body, the relief of pain and suffering, by medication, surgical interventions, psychotherapy, *etc.* At this level, the patient's good depends on the right use of the physician's knowledge and skill, those which are intrinsically part of the medical *tekné.*

But the medical good must be brought into proper relationship with the other levels of the patient's good. Otherwise, it may become harmful. What is medically "good" simply on grounds of physiological effectiveness may not be "good," if it violates higher levels of good, like the patient's good as he perceives that good. This perceived good is the second level of patient good.

THE PATIENT'S PERCEPTION OF THE GOOD

The medical good serves the many, complex facets of what the patient perceives as his own good. Here, we are concerned with the patient's personal preferences, choices, and values, and the kind of life he wants to live, the balance he strikes between the benefit and burdens of the proposed intervention. These qualities and values are unique for each patient and cannot be defined by the physician, the family, or anyone else. They are determined by the inter-relationships between and among age, gender, station in life, occupation, *etc.* To serve the general good of the patient, the medical good must be placed within the context of *this* patient's life-plans.

THE GOOD FOR HUMANS

Medical good and the patient's perception of the good life must be related to the good for humans as humans. This was the good Aristotle and Aquinas sought to define. At this level, we are concerned with the good peculiar to humans, like preservation of dignity of the human person, respect for his rationality as a creature who is an end in himself and not a mere means, whose value is inherent and not determined by wealth, education, position in life, *etc.* The patient is a fellow human with the physician to whom he is bound by solidarity and mutual respect.

It is at this level that some of the familiar principles of medical ethics are philosophically rooted, like autonomy, beneficence, non-maleficence, and justice. In American bioethics, these principles are taken to be *prima facie* principles rather vaguely grounded in a "common morality." This suggests that they could be changed if the common morality were to change them. For others, this is not the case. The good for humans is not subject to social construction. It lies within human nature and is a requirement of the natural law.

In the clinical encounter, the medical good and the personal good must, in their turn, be consistent with, and protect, the good for human beings. Physicians who ignore the patient's notion violate the good of the patient to self-determining rational being. Denial of care to the poor violates their dignity and value as human beings. Devaluing the lives of the handicapped does the same. Putting patients at risks that outweigh potential benefits, even with patient consent, violates the duty of beneficence and avoidance of evil.

SPIRITUAL GOOD

The highest level of good which must be served in the clinical encounter is the good of the patient as a spiritual being, *i. e.,* as one who, in his own way, acknowledges some end to life beyond the material. This may, or may not, be expressed in religious terms. But all, except the most absolute mechanistic materialists, acknowledge a realm of "spirit," however differently they may define it.

This realm of spirit gives ultimate meaning to human lives. It is that for which humans will often make the greatest sacrifices of other good things. For many people, the realm of the spirit is religion. They would be guided by a set of specific beliefs or doctrines that carry ultimate weight in every kind of decision. From the perspective of natural law, the spiritual destiny of man is his highest and ultimate good. Indeed, the dictates of the natural law are, themselves, that portion of divine law ascertainable by human reason.

Whatever the origin and content of one's spiritual beliefs, the three lower levels of good I have described must accommodate to the spiritual good. For example, blood transfusion might be medically "indicated" for the Jehovah's Witness, abortion of a genetically impaired fetus for a Catholic, or discontinuance of life support for an Orthodox Jew. But in these cases, the mere medical good could never be a healing act since it would violate the patient's highest good. Similarly, the Muslim, the Buddhist, the Hindu, or the humanist patient has his own spiritual good which must be encompassed within a clinical decision if it is to serve the "good" of the patient.

SOME COMPLEXITIES

In many clinical encounters it may not be possible to assess each of the four levels of patient good and establish the order of priorities among them. This is the case with infants, children below the age for responsible decision-making, the retarded intellectually, the elderly, or those in permanent vegetative states. In these circumstances knowledge about the patient's personal preferences or spiritual beliefs may be lacking. Yet clinical ethics imposes the duty to come as close as circumstances permit to an estimate of the patient's good as a whole.

In such cases, two levels of the good are still accessible - i.e. the medical good, and the good of the patient as a human being. In the case of infants it is impossible to know about preferences, etc. Parents, surrogates, and others must be relied on to represent the particularities of the other two levels from prior knowledge of the patient. How these are best balanced against medical good is a matter for more extended analysis than permissible here. Suffice it to say, that surrogates to be morally valid representatives must be without significant conflicts of interest. In the end the physician is still responsible for what is done and must therefore remain a guardian of the welfare of the most vulnerable patients.

In a pluralist society another issue centers on the degree to which a particular patient's preferences, world-views, and religious practices impinge on the physicians own beliefs about what is good for the patient. While I have emphasized the primacy of the good of the patient, the good as perceived of the patient and the physician himself must be respected as well. The dictum of the primacy of the good of the patient, does not eradicate the physicians judgment of what is good medicine, what he thinks of the human life ìssues and the spiritual destiny of himself and his patient.

In short the end of the clinical encounter is a subsidiary good, a part of the more complex realm of the good for humans. The physician therefore cannot be used as a mere means any more than the patient. The physician is entitled to respect for his autonomy, religious beliefs and scientific integrity. The complexity of resolution of conflicts between and among patients, physicians, families and other health workers is a growing problem in clinical ethics. But the existence of complexities cannot be used to justify a utilitarian, legalistic or libertarian definition of the ends of medicine or the physician's or patient's good.

Similar difficulties are encountered in the interpretation of clinical futility - which is a central factor in deciding the medical good. Clinical futility in any patient can be defined as the balancing of a relationship between the effectiveness, benefits and burdens of intervention. But this calculus may result in different or opposing conclusions about whether it is morally permissible to discontinue treatment

From the moral point of view there is no obligation to provide futile, or disproportionately burdensome treatments. This would be to do harm, not good. Yet some patients families want "everything" done, including cardiopulmonary resuscitation even when the data show it would be unsuccessful. These and many other conflicts in the interpretation of the "good" of the patient accompany the clinical encounter today. They promise to become ubiquitous in our morally polyglot society.

All of these complexities emphasize the need for greater clarity in the ethics of the process of decision-making. They do not vitiate or trivialize the importance of a definition of the ends of medicine. Indeed without some clarity on the ends of medicine and the good of patients as the end of medicine, the process of ethical decision-making will be even more difficult to establish.

This paper confines itself to a definition of the ends of medicine and the good of the patient. It does not engage the issues of the ends of investigative medicine or social medicine. These are realms in which medical knowledge and physicians are involved for ends different from the immediate ends of clinical encounters. However, in research and social medicine, the quadripartite conception of the good that I have developed in this essay must be preserved. The good is of a piece, and its unity must be intact even though, in practice, there may be tensions between and among its components.

IMPLICATIONS FOR THE OTHER HELPING PROFESSIONS

The conceptual schema proposed in this essay to redefine the ends of medicine in terms of the good of the patient has applicability beyond medicine. With proper redefinition of the ends peculiar to each profession, this schema can be used to define the good of the lawyer's client, the teacher's student, and the minister's penitent or parishioner. As with medicine, the ends of these other helping professions are linked to a particular activity specific to each profession.

Those who seek out these professionals share a certain common phenomenological ground. They all deal with human in compromised existential states. The persons they serve are dependent, anxious, in distress, and lacking something essential to human flourishing. That lack in medicine is health; in law, justice; in education, knowledge; and in ministry, union with God. Humans in these compromised existential states are eminently vulnerable and exploitable. Persons in that state are invited to trust the professional and, indeed, *must* trust him in order to be helped or healed. In each instance, the untrustworthy professional could exploit the patient's vulnerability for personal power, profit, or prestige. In each case, the character of the professional is the final safeguard. In each case, the end of professional activity is the good of the person in need of help.

As with the medical relationship, the "good" in each of the other three helping professions is a quadripartite concept: (1) the level of technical good, (2) the good perceived by the person served, (3) the good for the person as a (flourishing) human being, and (4) the spiritual good. Each profession operates most directly on one or other of the four levels. But, regardless of its specific focus, each profession must also attend the totality of the good of the person served by that profession.

For example, the lawyer focuses on obtaining justice for his client. Justice is a good of the client as a human being necessary to fulfillment of his human nature. But he cannot attain that end unless he is also concerned with the first level of the good - the legal good, *i. e.,* he must be competent in legal practice. He must be fully competent in legal procedures and in those techniques necessary to press his case in court, negotiations, or depositions. These are necessary for a right verdict but not entirely sufficient for a good verdict, for the lawyer must also be aware of the other levels of the client's values.

Thus, at the second level: What does the client deem justice to be in his case? To what degree does he wish to risk gain or loss, to settle out of court, to demand retribution? At the third level, the lawyer's success or failure is attendant on the degree to which he can gain for his client the human right of justice, freedom, vindication, or, if his client is guilty, a fair sentence. Finally, at the fourth level, to the extent that his client's religious or spiritual beliefs shape his plea for justice, they must be taken into account. For example, the client may be willing to forego certain of his claims in the name of charity for his opponent. Thus, as in medicine, all four levels of the good must be factored into the outcome or ends of the morally valid lawyer-client relationship.

Teaching may be similarly treated. As the possessor of knowledge and skill, the teacher's major emphasis will be at level three. Knowledge and truth are goods essential to human flourishing and fulfillment. To help others to achieve those goods, the teacher must, at the first level, possess the knowledge and skill he purports to teach. He must have mastery of the teaching methods, sources, and technical apparatus without which the end of knowledge transmission cannot be attained. At the second level, teachers must also adapt, to some significant degree, to the interests, learning modes, work habits, and preferences of the student. In the interests of the good of the student, teachers also may have to modify, restrain, or replace values that interfere with the end of learning. At the third level, that of the good for humans, teachers are required to respect their students with the dignity owed to persons, to treat them fairly, honestly, *etc.* Finally, at the fourth level, spiritual beliefs must be respected, allowed to flourish, and integrated with the more technical or academic dimensions of the student's education.

Finally, the priest-petitioner relationship(I am using the priest-penitent's relationship as an example of the relationship of the minister as spiritual healer. There is no conventional noun parallel to doctor-patient, lawyer-client, or teacher-student relationship. Priest-penitent stands here as symbol for the personal moral relationship of any minister of religion to the person who seeks his help) has its moral dimension most specifically at level four, the level of the spiritual good. This is what the penitent seeks from the minister: counsel on his relationship with God; how to be reconciled with God after sin; how to grow in the spiritual life; how to decide moral questions in light of revelation or church teaching; how to adapt to death, hardship, *etc.* in accordance with Divine Will. To attend to the spiritual good takes precedence over the other levels of good. But the other levels cannot, by that fact, be ignored since they are part of the integral good of the person served.

At the first level, therefore, priests must be skilled in the ends of the activities specific to the priestly vocation, *i.e.*, mastery of the theological or pastoral skills required to make their counsel serve the good of the penitent. At the second level, they must take into account the patient's unique values, the uniqueness of his spiritual predicament, his station in life, preferences for spiritual charisma, styles of prayer, life situation, *etc.* At the third level, the priest must protect the good of the patient as a human being, *i.e.,* maintain the seal of the confessional, help the penitent to integrate his spiritual and his temporal good, appreciate his dignity as a child of God, *etc.*

In each profession, the four components of the patient's good are arranged in lexical and hierarchical order. The spiritual good takes precedence over all, followed, in descending order, by the good for humans, the personal evaluation of the good, and, at the lowest level, the technical good specific to each profession. Moral decisions in the course of professional activities are "right" if they conform to the *tekné* of each profession at the first level. But to be "good," they must conform at the other three levels as well.

PATIENT GOOD, VIRTUES, AND PRINCIPLES

Thus, there is another facet of professional conduct common to all helping professions, the fusion of a technically night and morally good decision and act on behalf of a vulnerable human being. This fusion requires certain virtues or character traits in the practitioners of all the healing professions. David Thomasma and I have examined the question of virtue in general and specific virtues in the health professions elsewhere good of the person (E. Pellegrino and D. Thomasma, 1993, 1996), so I will only enumerate the professional virtues here.

I understand virtues as Aristotle did when he said " ... the virtue of man also will be the state of character which makes a man good and makes him do his own work well" (Nicomachean Ethics 1106a22-25). In the professions, the virtues are, therefore, those traits that dispose the professional - doctor, lawyer, minister, teacher - habitually to be a good person and do his work well, that is to say, to achieve the ends to which his work is directed. To be "good," the professional must predictably exhibit certain dispositions, those which enable his "work to be done well."

The virtues of the professions are entailed by the end of professional activity, *i. e.*, the good of the person served. Some of the most essential virtues would be: (1) fidelity to trust, which is ineradicable in healing relationships; (2) some suppression of self-interest, since the person served is in a vulnerable state and dependent on the power of the professional; (3) intellectual honesty, since professional practice beyond one's expertise is injurious; (4) compassion, since understanding and feeling something of the unique predicament of the person's need is essential to healing; (5) courage to pursue the good in the face of today's commercialization, depersonalization, and industrialization of professional life; and (6) prudence in every act so that the measures chosen are best suited to the technical and moral good of the person served.

Finally, a clear perception of the good of the patient, client, student, or penitent provides a moral grounding for the commonly employed "principles" of medical ethics. The good of the patient is a stronger foundation than the "common morality" adduced these days (Beauchamp and Childress). After all, the common morality is a social construction subject to change and unpredictability.

The good of the person served is linked ontologically to the end of the professional activity. It is not subject to change at will. With the good as the end of professional activity, autonomy becomes essential since to violate autonomy is to violate the dignity and humanity of the person. Justice is tied to the good for humans *qua* humans and is critical to every professional act for individuals and human society. Beneficence becomes the *primum principium* of all ethics, professional as well as general, since its end is doing good.

In sum then, the four helping and healing professions have a common end - the good of the vunlerable persons in need of professional help and expertise. Each profession deals with humans in vulnerable states; each confronts the most personal, intimate recesses of the lives of other humans; each is permitted access to the inner life of another human being; each promises to help and invites trust; each is judged by the degree to which the good of the person served is attained by their professional activities. Although each profession functions most directly at one or another of the four levels of human good, each professional also must, in its own way, serve the other levels as well.

We are speaking, here, of a common devotion shared by each helping profession to the good of the persons served. That good is the end of all professional activity. Subsidiary ends may be different for each profession depending upon which level of the good is its major focus. Whatever that subsidiary end, however, certain virtues and principles are necessary, and they are defined in terms of the being served.

CONCLUSION

Do good and avoid evil is the *primum principium* of all ethics. All ethical systems, medical ethics included, must begin with this dictum, which means that the good must be the focal point and the end of any theory or professional action claiming to be morally justifiable.

Since the Hippocratic Oath, the good of the patient has been the end of medical activity in clinical medicine. This paper attempts to define the good of the patient in concrete terms related to the phenomenology of the clinical encounter. The good of the patients is found to be a quadripartite good, a complex inter-relationship between medical, personal, human, and spiritual good, hierarchically arranged. This concept generates the duties of the clinician. The complexities of its application in medical practice are described. A theory of the good of the patient also has applicability for the ethics of the other healing and helping professions and the virtues and principles pertinent to their practitioners as well.

Edmund Pellegrino, John Carroll Professor of Medicine and Medical Ethics, Center for Clinical Bioethics, Georgetown University, Washington, D. C., U.S.A.

REFERENCES

Aristotle (1984), *Nicomachean Ethics.* In The Complete Works of Aristotle, The Revised Oxford Translation ed. Jonathan Bames, New Jersey: Princeton University Press, 1984.

Beauchamp T.and Childress J.F.(1994), *Principles of Biomedical Ethics.* 4 ed. New York: Oxford University Press.

Montague B. (1996), *The Quest for Moral Foundations: An Introduction to Ethics.* Washington, D.C. : Georgetown University Press.

Conway P.(1996), *Metaphysics of Aquinas, A Summary of Aquinas' Exposition of Aristotle's Metaphysics.* Edito da Mary Michael Spangler, Lanham: University Press of America.

Cooper, J. M. (1975), *Reason and Human Good in Aristotle.* Cambridge: Harvard University Press.

Finnis J. (1984), "Practical Reasoning, Human Goods and the End of Man," *Proceedings of the American* Catholic *Philosophical Association,* pp. 23 -36.

Douglas F. (1984), "On Two Meanings of Good and the Foundation of Ethics in Aristotle and St. Thomas," *Proceedings of the American Catholic Philosophical Associations,* pp. 56.

Hanson M. J. and Callahan D.(1999), *The Goals of Medicine*, The Forgotten Issues in HealthCare Reform, Washington, D.C.: Georgetown University Press.

W.E.R. Hardie W. E. R. (1980), *Aristotle's Ethical Theory,* Oxford: Clarendon Press.

Jonas H. (1984), *The Imperative of Responsibility in Search of an Ethics for the Technologic Age,* Chicago: University of Chicago Press, pp. 51-78.

Kraut R. (1989), *Aristotle on the Human Good.* New Jersey: Princeton University Press.

Mclnemy R. (1982), *Ethica Thomistica*, Washington, D.C. : Catholic University of America Press.

Mclntyre A.(1966), *A Short History of Medical Ethics,* New York: Macmillan.

Mclntyre A. (1990), *Three Rival Versions of Moral Inquire*, Notre Dame, Indiana: University of Notre Dame Press.

Nordin I. (1999), "The Limits of Medical Practice," *Theoretical Medicine and Bioethics,* Vol 20 (2), pp. 105-123.

Ellen Frankel P. , Miller Jr., Jeffrey P. (1992), eds. *The Good Life and the Human Good.* Cambridge, Mass. : Cambridge University Press.

Pellegrino, E.D. *The Ends of Social Medicine and Investigative Medicine*. In preparation.

Pellegrino, E. D. and. Thomasma D. (1993), *The Virtues* in *Medical Practice,*New York: Oxford University Press.

Pellegrino E. D. and Thomasma D. (1996), *The Christian Virtues* in *Medical Practice,* Washington, D.C.: Georgetown University Press.

Plato (1937), 'The Euthyphro," in *The Dialogues of Plato.* New York: Random House, p. 391.

DIEGO GRACIA

THE FOUNDATION OF MEDICAL ETHICS IN THE DEMOCRATIC EVOLUTION OF MODERN SOCIETY

INTRODUCTION

The moral world of modern times is significantly different from that of previous centuries. The idea that all human beings are entitled with a set of basic rights, which enable them to be treated as equal, and respected also in their religious, ethical or political diversity, is basically modem. During centuries the opinion was exactly the opposite: human beings are naturally different, and only the best of them can and must define what is right or correct in religion, morality or politics. The classic description of these natural and unavoidable differences was made by Plato, when in the *Republic* he differentiates the three types of human beings:

> "The god who made you mixed some gold into those who are adequately equipped to rule, because they are most valuable. He put silver in those who are auxiliaries and iron and bronze in the farmers and other craftsmen."[1]

Only the golden citizens "are adequately equipped to rule," political and also morally. All other members of the city should only be "supporters of the guardians' convictions"[2]. Plato is talking about the "perfect guardians" *(phylakas panteleis),* the only ones who can be called "rulers" of the city. They are natural rulers of the others because they are "perfect" *(pantelés),* because the *télos* of nature is perfectly realised in them. Therefore, they are not only good, but "the best". This is the meaning of the Greek word *àristos,* as superlative of *agath6s,* "good." This idea of government, not only of political government, but also of moral one, is "aristocratic." Only the best can know clearly the rules that must govern the moral life of people. And, as Plato says, to fulfil this goal the use of compulsion is moral when necessary:

> "The law's concern [is] to contrive to spread happiness throughout the city by bringing the citizens into harmony with each other through persuasion or compulsion and by making them share with each other the benefits that each class can confer on the community."[3]

Opposite to this view is the modern one, in which the ideals of happiness must be defined freely by individuals and social groups, and the use of coercion in order to impose a particular idea of perfection and happiness over the others, is prohibited, or at least considered immoral. The first task of the new morality is respecting the diversity of ideas of good and projects of perfection and happiness. This is the origin of the so-called "pluralism." Morality is not directly related with some specific contents; it is a structural or formal condition of all human beings. Therefore, all they must be respected, also when they are defending contents we do not consider good, or, at least, the best. This is the first principle of a "non-aristocratic" or

C. Viafora (Ed.), Clinical Bioethics. A Search for the Foundations, 33-40.

"democratic" morality. And the second is that public rules should be established in a process of deliberation between all human beings affected by the norm. In the definition of public rules nobody should have priority over the others. A public rule can only be considered fair when all these affected by the norm can accept it without coercion, through a process of rational and free deliberation.

In any case, ethics is closely related with politics. At the beginning of the *Nicomachean Ethics,* Aristotle said that ethics is "a sort of political science."[4] The Greek word *politiké* he uses can be translated by "politica" or also by "social." Ethics is a social enterprise; and bioethics also. And the question is in which direction we consider it must be developed, in an aristocratic way, or, on the contrary, in another democratic; or, also, if we can imagine a mixture of both, in a way in which democracy and aristocracy could not only coexist, but also improve one another their mutual performances. This is my point of view, that bioethics can not renounce to both, and that therefore it must articulate aristocracy and democracy in two levels, looking for no less than aristocracy in the "maximalist" level of the private ethics, and defending democracy in the "minimalist" level of the public one.

SOME DEFINITIONS: ARISTOCRATIC, DEMOCRATIC, AND STRATEGIC ETHICS

"Aristocratic ethics" can be defined as that which affirms or defends that some persons are entitled by someone (God, reason, nature, etc.) to define what things are good or bad, and that therefore they must naturally lead the others, whose first and most important moral virtue must be obedience. Things are good or bad by themselves, and therefore human beings should be only respected when they were doing rightful acts. Respect is not a direct propriety of human beings but of good actions. When people do not act correctly, they can not ask for respect.

On the contrary, "Democratic ethics" can be conceived as the ethics which considers that human beings are moral agents and that they are entitled of dignity due to their human condition, and not to the ideas or values they defend, or of the acts they make. Human beings, therefore, must be respected also when they do not agree with our political, cultural, moral, or religious beliefs. Here pluralism and dissent is possible; and not only possible but also necessary.

Both types of ethics are different from another that today is generally known as "Strategic ethics." This third type is the most frequent in the political life of our societies, and this permits understanding the great public discredit of our political institutions, specially from the moral point of view. People are convinced that politicians are more interested in the use of the public power for particular purposes than in the search of the common good and the public interest. The great figure is neither Plato nor Kant or Rawls but Machiavelli.

TOWARDS A DEMOCRATIC ETHICS

When saying "democratic ethics", we are giving to "ethics" a substantive role, and using the word "democratic" as qualifier. We are looking, therefore, to an ethics in which all human beings, due to their condition of human beings, will be taken into account. Nobody will be rejected or marginalized. All of them are entitled with dignity, as Kant said, and therefore are "ends by themselves" and not only means. The empirical differences of all type, natural or cultural, material or ideological, will not be used as reasons for exclusion. Here the universalization principle is unavoidable and compulsory. The canon of morality can not be considered correct or right with less than the inclusion of all human beings.

Kant expressed this canon with the different formula of the categorical imperative. One says as follows:

> "Act only according to that maxim whereby you can at the same time will that it should become a universal law."

Another formula is the following:

> "Act in such a way that you treat humanity, whether in your own person or in the person of another, always at the same time as an end and never simply as a mean".[5]

UNIVERSALIZATION AND INTERSUBJECTIVITY

The Neokantianisms of the second half of our Century have assumed without problems the Kantian universalization principle, but formulating it in an intersubjetive way, instead of the purely subjective proposed by Kant. The canon of morality must include all human beings, making them present in the process of argumentation and giving them the possibility of expressing and defending their own points of view. Neokantian ethics is, therefore, dialogic and deliberational, instead of monologic and apodictic. The works of K.O. Apel and J. Habermas have been essential in order to understand that norms and decisions can only be considered correct when they can be assumed freely and reasonably by all the participants. This is the principle called by Habermas (U), or principle of universalization:

> "For a norm to be valid, the consequences and side effects that its *general* observance can be expected to have for the satisfaction of the particular interests of *each* person affected must be such that *all* affected can accept them freely."[6]

The realisation of this principle is only possible when two conditions are fulfilled: first, "participation" in a true discourse of all those affected by the norm, and second, rational or reasonable "argumentation," because in other case it could not be considered a true "dialogue." Participation of all is a necessary condition, because in other case the (U) principle could not be actualised. That is why Habermas adds to the "universalization principle" another one, which he calls "principle of discourse ethics" or (D). This principle postulates the following:

> "Every valid norm would meet with the approval of all concerned if they could take part in a practical discourse."[7]

The other condition is the possibility of "argumentation" within true "dialogue." What is true dialogue? What kind of discourse and argumentation are adequate to true dialogue? Is true dialogue possible in empirical conditions? How can the ideal conditions established by the principles (U) and (D) be implemented in the real world? Apel answered this question distinguishing two parts in his discourse ethics, the Part one, ideal or canonical, and the Part two, actual or empirical. Factual conditions can make impossible the fulfilment of all the exigencies implied by the ideal community of communication in the real world, at a certain moment. Therefore, exceptions are possible. The only one unavoidable moral duty is the realisation of the ideal conditions in the real world in the shortest possible period of time.

The way followed by Habermas has been quite different. He thinks that the ideal conditions must be realised in the present world, throughout the transformation of the social and political structures. Habermas is not only a philosopher, but also a social and political thinker. And his thesis is that moral philosophy is necessarily engaged in the realisation of the ideal criteria; therefore, in law and politics. It is impossible for a cogent moral philosophy avoiding the fields of norms and facts. Consequently, ethics must influence Law and Politics. This is the argument of his book *Between Facts and Norms: Contributions to a Discourse Theory of Law and Democracy.*[8] The political problem is how to make suitable facts with norms. And the answer of Habermas is to transform our representative democracies in other more participative, and specially more deliberative. In other case, it is quite impossible to speak neither of a "democratic ethics", nor of an "ethical democracy."

DELIBERATION AS PROCEDURE

The example of Habermas is interesting, because he not only has done the effort of understanding ethically democracy and democratically ethics, but he also has proposed a method or procedure to reach this difficult task. And the procedure is "deliberation." Habermas took the idea of deliberation from some theorists of political philosophy, specially from Joshua Cohen, who in 1989 wrote: "The notion of a deliberative democracy is rooted in the intuitive ideal of a democratic association in which the justification of the terms and conditions of association proceeds through public argument and reasoning among equal citizens. Citizens in such an order share a commitment to the resolution of problems of collective choice through public reasoning, and regard their basic institutions as legitimate in so far as they establish the framework for free public deliberation."[9] This is the adequate procedure of analysing practical questions, viz. social, political and ethical, and therefore the way of managing the interpersonal relationships. According to Habermas, the deliberative democracy of Cohen is based in the following postulates:

> "(a) Processes of deliberation take place in argumentative form, that is, through the regulated exchange of information and reasons among parties who introduce and critically test proposals. (b) Deliberations are inclusive and public. No one may be excluded in principle; all of those who are possibly affected by the decisions have equal chances to enter and take part. (c) Deliberations are free of any external coercion. The participants are sovereign insofar as they are bound only by the presuppositions of communication and rules of argumentation. (d) Deliberations are free of any internal coercion that could detract from the equality of the participants. Each has an equal opportunity to be heard, to introduce topics, to make contributions, to suggest and criticise proposals. The taking of yes/no positions is motivated solely by the unforced force of the better argument. "[10]

This kind of deliberation is necessarily ideal, because the symmetry between human beings will never be completely actual; also because some internal and external constraints are inevitable, etc. It is therefore necessary to define deliberation in a more realistic way, as Aristotle conceived it at the very beginning. Deliberation is before all the belief in the incomnensurability of reality, and therefore in the need of including all the different approaches and perspectives in order to enrich the discussion and the comprehension of things and facts. Deliberation begins with the relativisation of the one's own perspective about phenomena, and the capacity of taking into account the perspectives of the others, discussing rationally their points of view, and modifying progressively one's own view through the process. Deliberation is a way of knowledge, because during it everybody engaged in the process is continuously testing and changing its own opinions, peacefully, without constraints. A process of deliberation has worked well when the points of departure of all the participants are different to the ones reached during the dialogue and defended by them at the end of the process. And the frequent consequence of deliberation is that the final solution taken in common did not coincide generally with the positions defended by any participant at the beginning of the process.

Deliberating is an art, based in the mutual respect, certain amount of intellectual humility, and the desire of enriching one's own comprehension of facts, hearing and interchanging opinions and arguments with the others engaged in the process. Deliberation is a way of critical and public analysis of one's own points of view. It requires certain knowledge, but specially certain skills and after all some attitudes and character. A person with grave psychological constraints, like unconscious fears and rigid prejudices, without the capacity of analysing and verbalising them peacefully and without anxiety, do have a more or less reduced capability of intervening actively in a process of deliberation. Experience shows that the process in itself works as educational, improving the performances of the persons involved. It can be said that nobody knows how to deliberate naturally. Deliberation is not a natural behaviour but a moral one. Naturally everyone thinks that he has the truth, and that all those who maintain different opinions or beliefs are either wrong minded or bad persons. Deliberating is a process of selfeducation. Perhaps it is also a process of autoanalysis, and at some extent also a therapy. Socrates spent his whole life educating young people through deliberation, or in the deliberation process. Socrates did not answer questions, but helped people finding their own answer to the questions.

DELIBERATION AND BIOETHICS

Aristotle taught deliberation as the main procedure of ethics. Practical reason is deliberative. Every individual deliberates with his own when he takes a personal decision. And everyone must deliberate with the others when there are two or more persons affected by the decision. Therefore, deliberation is the main procedure of discussing questions and taking decisions interpersonally. The doctor-patient relationship is not merely a contractual relationship, or a process of negotiation, as it is frequently affirmed. The doctor-patient relationship is a common process of deliberation.[11] The same can be said from the Hospital Ethics Committees. Its rationality can neither be dilemmatic, nor strategic, but deliberative. This is, perhaps, one of the most serious problems they suffer today, that many of them are working in a completely inappropriate way. Similarly, the work of all other private or public

Commissions or Committees must be deliberative. The success of the National Commission and of the President's Commission was not due to the reasons adduced by some of its members, but to the use of a paradigmatic deliberative procedure.[12]

And, finally, health care as a social and political institution should be also analysed within the framework of the so-called deliberative politics, or deliberative democracy. Oregon was a first attempt towards this direction.[13] Daniel Callahan has proposed over the last years to face some of the most cogent bioethical problems, as stopping the use of life sustaining treatment in elderly people and distributing the scarce economical resources, this way.[14] The same can be said of other proposals, like those of Amy Guttmann.[15] Perhaps we are at the beginning of a big turn in bioethics, from a dilemmatic and decisionist mentality to other more problematic and deliberationist.

DELIBERATION AND POLITICS

From a deliberative point of view, ethics is necessarily bound with politics, exactly as Aristotle stated. In fact, the modem theory of deliberation has been more developed by political philosophers than for ethicists. Relationships between human beings should be based in deliberative procedures, in order to make them reasonable and moral. In political relationships, only the "strategic rationality" has room nowadays. And deliberation is perhaps the best remedy against strategic politics, looking for true moral compromises.

> "The process of reaching any compromise is the same for all its forms, including moral compromises: the give and take of discussion and debate. It is by virtue of this process that moral compromises are still compromises rather than mere coercion. However, moral compromises about deep conflicts are quite different from the standard cases of strategic bargaining or tradeoffs. Their structure is dialogical and hence requires some forum for public deliberation. Their aim or result is not the balancing of moral concessions from both sides but a change in the common framework for democratic deliberation. As in all compromises, the parties begin with opposing values and standards. However, here they do not merely 'split the difference' fairly, nor do they find some impartial third position. If standards of fairness and impartiality are at issue, the very procedures of compromise that appeal to them must be modified in public deliberation. In these cases of conflict, a compromise is formed as each party modifies his or her interpretation of the common framework, often modifying that framework itself in doing so."[16]

Bioethics deals necessarily with the values involved in health and disease, the life and the death of human beings. Therefore, bioethics is a process of deliberation about the individual and collective ends of human life. Consequently, it can not be constrained to the limits of Hospitals and Schools of Medicine. The goals of human life are primarily social and political. And because the ends of medicine are derived from these goals, it is necessary to conclude that bioethics is engaged inevitably in matters that occur out of the hospitals and out of the health care professions. Bioethics is not only a professional ethics, but also and principally a part of political philosophy. As Ezequiel Emanuel has pointed out,

> "to make clinical decisions for an 'individual' patient, the practising physician must specify and balance the ends of medicine. But this process occurs within a framework

> constructed from ethical conceptions that have been elaborated by political philosophy. Medical ethics is thus a subfield of political philosophy." [17]

It is not necessary to agree with the way in which Ezequiel Emanuel thinks of political philosophy, to accept his idea of bioethics as a social and political matter, and not only a professional one. Emanuel is a convinced "communitarianist," but some other important "contractarian" authors, like Habermas or Rawls, do think similarly. For instance, Rawls has stressed that the aim of his theory of justice is

> "to formulate an ideal constitution of public deliberation in matters of justice, a set of rules well-designed to bring to bear the greater knowledge and reasoning powers of the group" in the development of laws and norms.[18]

Deliberation is the method of practical reason. Therefore, it must be promoted when the values and ends of human life, individual and collectively, are at stake. And bioethics should work in this way.

CONCLUSION: BIOETHICS AS AN ETHICS OF RESPONSIBILITY

The expression "ethics of responsibility" *(Verantwortungsethik)* was coined by Max Weber, as the avoidance of two extreme positions, the ethics of conviction on one side, which decide the morality of actions only by their contrast with absolute and universal principles, and the ethics of interests on the other, which take into account only the weigh of consequences. Both extremes are, at some extent, "irresponsible." This is, perhaps, one of the most widespread convictions of the ethics of the Twentieth Century. Moral acts must take into account principles and consequences. Both dimensions are necessary and unavoidable in a Responsibility ethics. Human morality is needed of balancing principles and consequences. Principles are universal and *a priori,* whilst consequences are particular and *a posteriori.* Principles are inescapable, but the evaluation of consequences is also necessary. And the way of balancing both sides is deliberation. Its final goal is always the same, promoting not only moral Democracy but also moral Aristocracy or Excellence. In other words, the final goal of a true moral life is today democratising aristocracy, or aristocratising democracy. This is, perhaps, the best description we can give today of an "ethics of responsibility."

Diego Gracia, Professor of History of Medicine and Director of Department of Public Health and History of Science, Complutense University Madrid, Spain.

NOTES

[1] Plato, *Republic* 415a.
[2] Plato, *Republic* 414b.
[3] Plato, *Republic* 519e.

[4] Aristotle, *Nicomachean Ethics, 1,2:* 1094b 11.

[5] Kant I. (1995), *Grounding for the Metaphysics of Morals,* in: S.M. Cahn (Ed.), *Classics of Western Philosophy,* Indianapolis/Cambridge: Hackett Publishing Company, 4 th ed. , p. 1080 and 1085.

[6] Habermas J. (1990), *Moral Consciousness and Communicative Action,* Cambridge, Massachusetts: The MIT Press, pp. 120, 197.

[7] Habermas J., *Moral Consciousness and Communicative Action,* p. 12 1.

[8] Habermas J. (1996), *Between Facts and Norms: Contributions to a Discourse Theory of Law and Democracy,* Cambridge, Massachusetts: The MIT Press.

[9] Quoted by Habermas J., *Between Facts and Norms: Contributions to a Discourse Theory of Law and Democracy,* p. 305.

[10] Habermas J. , *Between Facts and Norms: Contributions to a Discourse Theory of Law and Democracy,* pp. 305-6.

[11] Emanuel E.J. , Emanuel L. L. (1992), *Four models of the physician-patient relationship,* JAMA, 267(16), pp. 2221-6.

[12] The "Committee on the Social and Ethical Impacts of Developments in Biomedicine," of the Institute of Medicine of the National Academy of Sciences, has analysed in a complete report the work made by the different American bioethical commissions. And the conclusion reached is that the goal of these commissions should be what they call "Bioethics Deliberation." Committee on the Social and Ethical Impacts of Developments in Biomedicine. *Society's Choices: Social and Ethical Decision Making in Biomedicine.* Washington, D.C.: National Academy Press, 1995.

[13] B. L. Hines,B. L. (1985), *Oregon and American Health Decisions: A Guidefor Community Action on Bioethical* Issues, Washington, D.C.: Department of Health and Human Services. Crawshaw R., Garland MJ. , Hines B., and Lobitz C. (1985), *Oregon Health Decisions: An Experiment with Informed Community Consent, Journal of the American Medical Association,* 254, pp. 3213-16. See Emanuel E. J. (1991), *The Ends of Human Life: Medical Ethics in a Liberal Polity,* Cambridge, Mass: Harvard University Press, pp. 210-12.

[14] Callahan D. (1987), *Setting Limits,* New York: Simon and Schuster. Callahan D. (1990), *What Kind of Life,* New York: Simon and Schuster.

[15] Gutmann A. and Thompson D. (1997), *Deliberating about Bioethics, Hastings Center Report* ; 27(3), pp.38-41.

[16] Bohman J. (1996), *Public Deliberation: Pluralism, Complexity, and Democracy.* Cambridge, Mass.: The MIT Press, p. 9 1.

[17] Emanuel E. J. , *The Ends of Human Life: Medical Ethics in a Liberal Polity,* p. 23.

[18] Rawls J. (1971), *A Theory of Justice,* Cambridge, Mass.: Harvard University Press, p. 359. Quoted by Emanuel E.J., *The Ends of Human Life: Medical Ethics in a Liberal Polity*, p. 148.

HENK TEN HAVE

A COMMUNITARIAN APPROACH TO CLINICAL BIOETHICS

THE SIGNIFICANCE OF PHILOSOPHICAL ANTHROPOLOGY

At the close of his famous historical overview of the philosophy of medicine, Szumowski has proposed a definition of the discipline, as well as a distinction between several subsections: logic and ethics, and perhaps epistemology, psychology, and medical metaphysics (Szumowski, 1949). It is curious that he did not explicitly refer to philosophical anthropology. On the one hand, it is obvious that the problem of man has been a continuous object of reflection during the long history of philosophy of medicine. In many varieties, and from different perspectives, medicine, and therefore also the philosophical self-understanding of medicine, has attempted to answer the basic question: What is man? Time and again, medicine, as Szumowski shows, has grappled with issues of organicism and mechanism, holism and reductionism, vitalism and materialism. Some understanding of man is crucial for the mission of medicine. Whether it is the view of the person as a whole or the view that man is a sophisticated collection of particular organs, some view of the human being is essential to demarcate and define the object as well as the methods of medicine as science and practice. Man himself plays a central role in medicine as an entity subject to and suffering from disease; the human being is the object of interest as well as the subject of study. As for example William Inlow (1946) has pointed out, when these truisms refer to essential characteristics of medicine, philosophy of medicine cannot be developed without philosophical anthropology. In a very basic respect, anthropology necessarily is a presupposition of philosophy of medicine. Anthropological questions have been pervasive throughout the history of this discipline, perhaps so pervasive that they did not need to be identified separately. Philosophy of medicine in this perspective can be regarded as an uninterrupted essay on man. One of the consequences of this view has been presented by Szumowski in an 1937 essay. In medical education, students need a coherent framework of knowledge, practice and values; they need to envisage medicine as an unity. But then medicine and its representatives need to proceed from a unified image of man; they cannot continue to divide the patient into a disease and a person, into a body and a psyche, into an entity as a whole and its constituent organic parts (Szumowski, 1937).

But, on the other hand, medicine itself can also be considered man's effort to understand himself. Medical science not only generates knowledge and facts about human nature, but also contributes to particular concepts of man. Medicine, perhaps more than the other sciences, helps to produce and construct such concepts. Ernst Cassirer has pointed out that in science man can do no more than to build up his own

C. Viafora (Ed.), Clinical Bioethics. A Search for the Foundations, 41-51.

universe. If that is true, then medicine adds something more: through contemporary medicine man can build up himself as part of that universe. Medical activities articulate what concepts and ideas of man contemporary human beings attempt to realize. Medicine then can be one of the preeminent examples to learn more about the philosophical ideas about the human person that prevail in present-day culture. To medicine, Cassirer's statement can be applied that ...

> "Man's outstanding characteristic, his distinguishing mark... is his work. It is this work, it is the system of human activities, which defines and determines the circle of 'humanity'" (Cassirer, 1976, p.68).

From this perspective of medicine as self-expression of the human being, it is urgent to reflect upon the images of the human being exemplified, constructed and realised through medical activities. Philosophical anthropology no longer can remain an implicit presupposition but needs to be developed into an explicit, articulated and analytic activity.

THE ANTHROPOLOGICAL PHASE OF PHILOSOPHY OF MEDICINE

This is exactly what the history of philosophy of medicine learns us. Particularly from 1870 there has been a rapid growth of medico-philosophical literature. It is argued that in the thematic development of philosophy of medicine since that renewed starting-point three traditions can be distinguished: an epistemological, anthropological, and ethical tradition (Ten Have, 1995). This distinction can help to clarify the continuity of the basic themes, preoccupations, and motivations within the discipline of philosophy of medicine. The few historical studies available, notably Szumowski's (1949), do not demonstrate any evolution of the discipline; one may easily have the impression that there has been a succession of philosophising doctors and a rather accidental accumulation of books and articles. Retrospectively, it might even appear that philosophy of medicine is a very recent affair and that the present preoccupation with bioethics shows a marked discontinuity with earlier efforts to philosophise about medicine. The present-day domination of bioethics is, however, continuous with the various traditions of philosophy of medicine in that it is moved by the same commitments and fundamental problems. This continuity is manifest, especially in relation to the tradition of anthropologically oriented medicine which flourished prior to the current interest in ethics.

Present-day interest in bioethics can be regarded as the latest phase of a long tradition of theoretical reflection upon medicine (Ten Have, 1997). A basic concern during this continuous reflective effort is the attempt to understand the identity of medicine: What is this human activity that is called "medicine"? How should it be interpreted? What characteristics should be used to define it? In the first phase of modern medical philosophy the identity of medicine is described in epistemological terms. Medicine is characterized as a natural science. In this scientific conception of medicine, the artistic element, the art of medicine, is eliminated. But, at the same time, it is understood that the unity and coherence of medicine were endangered through the successes of its scientific approach. In the philosophical literature, two problems were identified: first, medical knowledge is fragmented and medical practice one-sided because of specialisation; second, the patient as the object of medicine is no longer adequately addressed since the conceptual tools of medicine are insufficient and too simple. Solution have been sought by proposing more rigorous methodologies, synthesis of medical knowledge in grand theories, and re-interpreting medicine as an art.

The interpretation of medicine as an art evolved into a new conception of medicine as anthropological science, - influential from 1930 until 1960, particularly in Germany and the Netherlands. Physicians and philosophers of medicine such as Buytendijk, von Weizsäcker, von Gebsattel, Plügge and Christian sought to redefine and re-interpret medicine as a science of man, applying and developing ideas from several contemporary philosophical schools, particularly phenomenology, existentialism, and philosophical anthropology. What is important in this conception is the tendency to idealise the doctor-patient relationship. The subject should be re-introduced into medicine; that implied acknowledgement of the subjectivity of the knowing and acting subject (the physician) but also that of the object (the patient). Medicine is considered a unique profession in systematically and methodically attending to the patient as an irreducible person.

Since the 1960s, this anthropological orientation was rapidly superseded by a growing interest in bioethics. However, there is a marked continuity between these two phases of philosophy of medicine. Through concentrating on the subjectivity of the patient, anthropological medicine paved the way for the subsequent ethical phase. It opened the moral dimension of medicine for public reflection, because it argued that medicine itself is a normative science of life. The current preoccupation with ethical problems is, in this view, not discontinuous with earlier efforts to philosophise about medicine. In a certain way, it shares the same commitments and fundamental problems as earlier phases, although with different concepts and vocabulary. Bioethics, therefore, is part of a long tradition of philosophical reflection in health care. However, what is new, is the tendency nowadays to phrase fundamental problems in the language of good or bad, right or wrong, acceptable or unacceptable. Furthermore, within such a traditional view, bioethics is not so much focused on solving these problems, rather than on clarifying their value-context, analysing, for example, the goals of medical practice and the subjectivity or personhood of the patient.

IMAGES OF THE HUMAN BEING IN MEDICINE

What the anthropological tradition in philosophy of medicine helps us to remember is that the labelling of problems and issues as "moral" requires a further clarification of the underlying images of the human being that guides our normative judgments. What exactly is the human being whose welfare, interests or values are harmed or benefitted through medicine? What kind of human being is presupposed in medical activities and the moral justifications of those activities?

It seems that the human condition of today is precisely characterised by a fragmentation of images of the human being. Zygmunt Bauman, scholar of Polish origin, and emeritus professor of sociology at the University of Leeds, distinguishes in one of his recent books four types of intertwining and interpenetrating images of postmodern man: the human person as the stroller, the vagabond, the tourist, and the player (Bauman, 1995). These images not only characterise various lifestyles but also lead to different self-interpretations of human beings, with far-reaching implications for human relationships and moral responsibility. Within a world that has replaced durable objects with disposable products, and guided by the determination to live one day at a time as the ideal of rational conduct, not committed to control the future but to refuse to mortgage it, postmodern life strategies are not focused on identity-building but on the avoidance of being fixed. When the horror of being bound and fixed is the most important determinant of the self-understanding of contemporary human beings, the scientific agenda of philosophical anthropology should focus on analysing the floating and discontinuous lifestyles, rather than identifying and delineating more or less cohesive and continuous ideas or concepts of man. Postmodern life is incoherent; every identified life strategy conveys only a part of the story, almost never integrating into a totality.

This diagnosis of the postmodern predicament, described and clarified by Bauman, provokes questions, such as: What are the implications of this diagnosis? How adequate or correct is it? What is the significance of an anthropology of fragmentary and discontinuous human life for philosophy of medicine? Furthermore, there also is the preliminary question how such diagnosis is possible. If fragmentation and discontinuity are the essential halmarks of human life, from what perspective are we able to perceive and recognise the postmodern image of the human being, we ourselves, as fragmentary and discontinuous?

In their writings, representatives of the anthropological movement in medicine tried to reflect upon human existence in its concrete specificity and ambiguity. Instead of starting from or working toward an ideal image of the human being, they attempted to identify what is anthropologically characteristic and common to human beings. But at the same time they were very much aware that any image was too abstract and "clean", because in everyday reality the specific individual was always changing, pluriform, and was not fully described by the designed image of a person. Given this theoretical point of view, advocates of anthropological medicine have not presented a clear-cut theory that has been defended and elaborated upon by all representatives.

What is important, however, is that they have helped to articulate and criticize the image of the human person, that underlies, justifies and stimulates much of everyday medicine: a universalistic and reductionistic image. In this image, human beings can be understood by analyzing and studying anatomical structures, physiologisch functions, pathological aberrations, biochemical complexities or genetical locations and dislocations. Such criticisms are not at all outdated or anachronistic; reductionist images tend to re-appear with every new and promising scientific development.

MAN AS MECHANISM

Health care in particular may illustrate the different ways in which we view, understand and approach human beings. Images of man involve metaphors and models; they lead to specific attitudes; they also pattern explanations and interpretations; they thus give specific orientations to care and assistance. Images of man not only express how human beings are, but also how they should be; these images are normative as well as descriptive. How we morally interpret and judge human activities and situations is determined by the particular image of the human being we have in mind.

The anthropological critique of medicine demonstrates that different images of man are influential in health care. The critique itself focuses on delineating and reviewing the most dominant image in modern medicine: man as mechanism. The mechanistic image of man underlying in a particular clinical and curative medicine is in fact the Cartesian heritage (Ten Have, 1987). In Descartes' philosophy, man is composed of two fundamentally different substances; he is a structure of *res cogitans*, a thinking entity, and *res extensa*, an extended entity. These substances, although they are related in the practice of daily life, are each other's opposites and not reducible to one another. Descartes' conception that matter, thus also the human body, is exclusively characterised by extensiveness in space, has important consequences. Matter has merely quantitative aspect, which may be described entirely through mathematics. Nature, therefore also the human body, may be explained in purely mechanical terms. Bodily processes may be examined and explained mathematically and physically. This conception of the body as an intricate machinery has become extremely influential; as an object that does not show a fundamental difference from other things, it can become the object of study in modern medicine conceiving itself as one of the natural sciences.

Considering the human body as part of material reality was a fruitful paradigm for modern medicine. Descartes used to compare the body with a clockwork. By analysing it one can find out the way it functions; physiology is the product of the constitution of the body and the mechanical interactions of the anatomical basic elements. Life and death, illness and health are manifestations of this self-maintaining entity the body is. With the same precision and certainty with which a broken watch may be mended, complaints of the mechanical body can also be cured, once the origin of the complaint has been diagnosed, *located* in the body. Medical cases are of a technical nature and in principle solvable. The more and deeper a physician's knowledge, the greater his technical and problems solving abilities. A long, health, and painless life is guaranteed as the fruits of medical science based on natural sciences. Descartes had great confidence in the progress of a medicine, based upon physics. By offering the method of scientific research (as the reduction of the complex to the simple), and the machine model of the body (as the reduction of the phenomenal to the mechanical), he shows the way to an effective scientific medicine, which could confer many practical benefits in mankind.

The problem is that the element of thought-construction, which charactizes Descartes' thinking with respect to the body, has disappeared. Speaking of the body *as* a machine is replaced by remonstrances *about* the machine of the body. The initial success of the mechanistic programme in biology, physiology, and pathology leads to the conception that with the interpretation of physical automatisms everything has been explained. The soul, separated from the body by Descartes, evaporates as the body's functioning becomes clearer. The effect is that in modern medicine the human being has been "naturalised"; he is considered part of nature rather than individual subject.

MAN AS PILGRIM

The anthropological critique focuses on rejection of the prevailing image of man, and on the introduction of a notion of medicine as science of the human person.

It is first of all argued that the dualistic ontology usually underlying medical theory and practice, subdividing human beings in a physical and mental compartment is not adequate when caring for ill people. Medicine, of course has profited enormously from this subdivision, but it has also restricted itself to the human body by studying and explaining the body's physico-chemical machinery. This approach has a tendency to reduce the human being to a specific animal species, and the human characteristics of the human body to its physical level of being. Anthropologically-oriented physicians argue that any demarcation between body and mind is artificial. Dualistic thinking, moreover, not only involves a reduced image of the human person, but it also has a more general tendency of making a stringent distinction between object and subject. Such thinking erroneously suggests that there is an objective, real world, independent from an isolated, individual subject. It also leads to an almost exclusive preference for the methods of the natural sciences in the context of health care. These methods are focussed on intervention, control and manipulation, introducing the technical point of view of the engineer into the domain of disease and suffering.

What is therefore needed is a more appropriate understanding of medicine as science of the human person. The point is not that the causal thinking and technical approach of the natural sciences should not be allowed in medicine; on the contrary, they are valuable, but they are not sufficient. Medical thinking and practising should not restrict itself to these scientific methods, because they cannot grasp what is essential to human beings. As a living organism, every person constitutes a whole, a meaningful entity, which is disconnected and disintegrated in abstract, analytical approaches. What we need, for example, is an anthropological physiology, not only explaining those aspects of living organisms that can be analysed from a mechanical point of view, but also clarifying the intrinsic meaningful connections and interrelations of bodily phenomena as specifically human, as well as the significance of human experience and conduct (Dekkers, 1995). If medicine wants to evolve into a science of the human person, it should overcome the distinction between the objective and subjective, introduing the subject into medicine itself. Buytendijk, one of the representatives of this anthropological traditions, summarises his position as follows: If medicine is not objective, it is impossible; if medicine is only an objective science, it is inhuman (Buytendijk, 1959).

What is true for the methodology of medicine, also applies to the medical understanding of disease. Focusing on the causal mechanism of disease, medicine cannot fully understand the ill person, because explaining disease also refers to the problem of the significance of a symptom, the meaning of a particular complaint (Welie, 1995). Science based medicine in fact hinders that insight that disease has meaning (Welie, 1994). For patients, relevant questions do not so much refer to the pathogenesis and pathophysiology of the disease, as to the anthropological query: "Why am I suffering here and now?". Being ill primarily is an existential category; only secundarily, we can make any differentiation between organic phenomena, subjective complaints and personal suffering. Being ill is a way of being a human person. From an anthropological perspective, human beings do not only have their lives, but also give expression to it; they do not merely have their bodies, but also are their bodies at the same time. From that same perspective, it is not only the case that persons have their disease, but they also make their disease as a response to his or her own individual existence. In this perspective, disease is not a negative event, a blind fate, waiting to be eliminated from the world by technological intervention; the important thing is what we make of it, whether we consider it as an occasion to reconsider and improve our life.

The problem with the anthropological critique of the mechanistic image of man is that the antidote, in other words the alternative image presented in anthropological medicine, is now itself questionable. For the anthropological critics human beings essentially are dynamic beings adapting to the world; they reconstruct and redefine their identity in order to keep it solid and stable; they consider their life as a project, an individual task to accomplish the goals that have been set, and to give meaning to the various stages and experiences. This is the image of human life as pilgrimage, described by Bauman (1994, 1995). For man as pilgrim, the true place is always some time, some distance away. What makes his life worthwhile is the proximity between the true world and this world; he is destined to be elsewhere; the pilgrim's life is worthwhile to the extent that he still journeys, but because the pilgrim actually wishes to attain his goal or destination, the closer he is to his goal, the more worthwhile his life. Because life has been transformed into a pilgrimage, it receives meaning. It is the destination that makes a whole out of the fragmentation, that lends continuity to the episodic. Human life therefore is a continuous story, and although it is an individual project, it is carried out in an orderly, determined, predictable world. The image of the human being as pilgrim, that has more or less implicitly motivated anthropological physicians for a long time, has specific characteristics (such as goal-directness, the deliberate choosing of successive steps in life, a slowly changing world requiring man adapting, unity and continuity of the life project) which do no longer apply, according to Bauman, to the postmodern situation.

MAN AS TOURIST

> "The life of modern man was frequently likened to the pilgrimage-through-time. The itinerary of pilgrims is drawn in advance by the destination they want to reach...and everything they do is calculated to bring them closer to the goal. The pilgrim is consistent in choosing every successive step, conscious that each step matters and the sequence cannot be reversed. Today's men and women can hardly treat their life as a pilgrimage, even if they wished to. (...) the life of the men and women of our [post-modern] times is more like that of the tourists-through-time: they cannot and would not decide in advance what places they would visit and what the sequence of stations would be; what they know for sure is that they will keep on the move, never sure whether the place they reach is their final destination" (Bauman, 1994, p.20).

The postmodern human being is characterized by "endemic non-finality" and "non-fixity" (Bauman, 1995, p.78 ff). There is no final state, no state of perfection which may be realised in search of improvement; life has only local and transitory achievements. The defining feature of postmodern existence is the fear of being fixed. Persons avoid fixation, commitment and stable relationships. They like to keep their options open. For them, the ultimate freedom is celebrated in the activity of zapping. The postmodern individual has ultimate control over the images he or she wants to produce, and can change them at will. Contingency, episodicity and fragmentation are the marks of human life. In a very basic sense, the image of the human being is merely and nothing more than an image; the image also is perpetually changing. Bauman therefore prefers to speak of life-styles and life strategies, rather than images of man. This world of life-styles seems no longer hospitable to pilgrims. The anthropological critique of modern medicine is not feasible anymore in this world of dissipative structures, that has lost its definiteness and continuity, this human reality that exists as a series of episodes or events without a past and with no consequences. The postmodern person is best characterised with the image of tourist. This person, in the words of Bauman, is

> "a conscious and systematic seeker of experience, of a new and different experience, of the experience of difference and novelty" (Bauman, 1995, p. 96).

For man as tourist the world is domesticated; it is obedient to his wishes, it is made to please. The most important thing is that the world should be structured by aesthetic criteria. For man as tourist it is, in the end, not clear where home is; having a home becomes a mere postulate. The tourist belongs nowhere but dreams of belonging. We do not need to agree with this diagnosis of postmodernity in order to recognise the consequences. The postmodern person is an interest-seeking subject. For him or her, the objects in the world are not relevant as entities in their own right. What matters is whether they are pleasing or not pleasing, satisfactory or unsatifactory. The tourist does not want to change the world, he wants to lead an enjoyable life. Reality can not challenge him; it does not provoke him towards rectification, improvement, transformation. Postmodern life strategies are furthermore focused on rendering human relations fragmentary and discontinuous. Doing so, they promote the distance between the individual and the other; the other person moreover is primarily considered as the object of aesthetic, not moral evaluation. The effect of these postmodern strategies of disengagement and commitment-avoidance is the suppression of the moral impulse. What transforms experiences into moral experiences has much to do with responsibility for the other, engagement in the fate of the other, and commitment to the other person's welfare.

COMMUNITARIAN ANTHROPOLOGY AND ETHICS

What postmodern literature makes clear is how the relationship between anthropology and ethics essentially is. Instead of the claim of the primacy of anthropology, with ethics based on anthropological theory and images of the human being, philosophy of medicine should also focus upon explicating the morality underlying anthropology itself. Ethical discourse can exemplify a particular image of man, but at the same time anthropological discourse itself is presupposing particular moral views. The question is not whether the image of the human being as a pilgrim or a tourist is an adequate description of the present-day situation. To verify or falsify such description is a matter of sophisticated sociological studies. Even with the most complicated study designs, however, it seems almost impossible to make the description into a general characterization of present-day human beings, since its truth-value is so much dependent upon the cultural and social circumstances; what is a true sociological representation of human beings in Texas most probably is not true for 'postmodern' human beings in Veneto, Catalonia or Gelderland. The more interesting question, however, is a normative one: What underlying notion of morality makes it possible to diagnose the human situation using such images? The possibility condition seems to be a basic understanding of morality, not as something we choose, but as a fundamental predicament we are already involved in before we even start to reflect upon it; such predicament, however, at the same is an anthropological characterisation of what is essential in human beings. Morality is choosing us, because we are primarily social beings. Ethical views are articulated and explicated because we are in a moral relationship with other human beings appealing to us. The face of the other makes us moral beings whether we like it or not, whether we choose to act accordingly or not. This inter-personal character of morality is the implicit philosophical background of the postmodern diagnosis of fragmented and discontinuous lifestyles. The diagnostic result is that moral views are considered objects of individual choice, items on the market place that we prefer or not, and exchange whenever we like. The outcome of the diagnosis, however, should not be confused with the possibility condition to make the diagnosis. Morality is a social affair. Its inter-personal character makes it possible to scrutinise and criticise individual moral choices. Tourism as an individual experience is only possible within a huge network of human cooperation. It annihilates itself when individual preferences destroy the quality of the network. The dialectic interaction of anthropology and ethics may help us to regain a view of man as social being, and therefore restore the idea of moral community. Anthropology proceeds from contextual practices which presuppose moral values. At first reading, this leads to the postmodern diagnosis that human beings are tourists, consumers, free and independent selves. But in a second-order reading, the process of diagnosing itself, the very condition making such diagnosis possible and understandable, shows the social embeddedness of individual life. It shows that our selves are constituted through the practices of the community. Cultural context and community are constitutive of the values and goals of individuals. Communal relatedness falsifies therefore the idea of the unencumbered self, the idea of self-ownership assuming that the individual as an entity exists prior to the ends which are affirmed by it. According to this second-order reflection, the idea that the self designs its life-project from a a-social or pre-social position, and subsequently participates in the community, is self-defeating. Without societal culture our potential for self-determination will remain empty. Therefore, the lesson of anthropology is that it explicates practices which define a given morality, showing that, regardless of particular and fragmented practices, the individual self is fundamentally dependent on commmunity.

At the same time, there is a third level of analysis. Although human beings are fundamentally dependent on community, man as moral being still has the potential to ethical reflection. Constituted as a self by societal culture, he at the same time can obtain some reflective distance, creating independency from the particular roles, goals and values that characterize the societal culture. Otherwise, the possibility of ethical reflection should only consists in explicating and articulating the values and goals prevailing in the societal culture, without the opportunity of critical disengagement.

It is precisely at this point where the philosophical struggle is located in recent works in social and political philosophy: if the postmodern human being is not an unencumbered individual or an autonomous consumer, but part of an encompassing community, how can he as a reflective moral being escape the universal dominance of the community to which he belongs? Michael Sandel (1996), for example, argues against the moral claims of "cosmopolitan citizenship" with its emphasis on the community of all human beings; in this view, universal identities take precedence over particular ones; ideally, the distinction between friends and strangers should be overcome in a truely ethical point of view. Sandel points out that most of the time we live our lives by smaller solidarities; we learn to love humanity not in general but through its particular expressions. What is morally relevant, therefore, is not the community of all human beings but the particular communities that locates us in the world. What is typical, furthermore, is that we live in a range of different communities, some overlapping, others contending. Moral reflection is necessary to decide which of one's identities is properly engaged. Postmodern individuals are not unencumbered by moral ties they have not chosen; but they are also not encumbered in a universal community with obvious encompassing loyalties. They are, however, citizens who can think and act as "multiply-situated selves". Ethical reflection is primarly needed to cope with the ambiguity associated with multiply-encumbered selves; it should give us

> "the capacity to negotiate our way among the sometimes overlapping, sometimes conflicting obligations that claim us, and to live with the tension to which multiple loyalties give rise" (Sandel, 1996, p. 350).

A similar argument is developed by Benjamin Barber (1995). Postmodern individuals are members of a world-wide community, so-called McWorld, the global theme park of MTV, Macintosh and McDonald's, a world tied together by communication, information, entertainment and commerce; in this world everyone is a consumer, defined by needs and wants. McWorld therefore is not really a community: the significant relations are exchange relationships among individual consumers and individual producers; society is privatized and commercialized. The question then is how to reconnect individuals with civil society and civic culture, as the middle ground between big government and the private sector. Interposed between the state and the market is where community exists, where we are more than clients or consumers, where we are public beings having regard for the general good, where we as citizens relate in the cooperative, noncoercive pursuit of public goods. Barber defines a citizen as

> "an individual who has acquired a public voice and understands himself to belong to a wider community, who sees herself as sharing goods with others" (Barber, 1995, p. 286).

But also Barber agrees that humankind depends for its liberty on variety and difference; we live in several spheres, in many-sectored civil society. Whereas market choices are private and speak about individualistic goals and individual preferences, citizens speak about the social consequences of their private choices; they speak the public language of the common good; but at the same time, this public language is multiple and heterogeneous; civil society has many narratives about the common good.

The *universal* human condition of existence as a communal-cultural being can only be realized in *particular* ways; the communitarian self is constituted by particular cultural characteristics. A richer medical ethics can result from taking seriously the basic idea of moral community, and concomitantly, the various narratives about the particularities of people as communal beings.

Henk Ten Have, Professor of Medical Ethics, Department of Ethics, Philosophy and History of Medicine, University Medical Center, Nijmegen, The Netherlands.

REFERENCES

Barber, B. R. (1995), *Jihad vs McWorld*, Ballantine Books, New York.

Bauman, Z. (1994), *Alone again. Ethics after certainty*, Demos, London.

Bauman, Z. (1995), *Life in fragments. Essays in postmodern morality*, Blackwell, Oxford.

Buytendijk, F.J.J. (1959), *De relatie arts-patiënt*, Ned Tijdschr Geneeskd 103, pp. 2504-8.

Cassirer, E. (1976), *An essay on man*, Yale University Press, New Haven and London (original 1944).

Dekkers, W.J.M.(1996), *F.J.J.Buytendijk's concept of an anthropological physiology*, Theoretical Medicine 16, pp. 15-39.

Have, H.A.M.J. ten. (1987), *Geneeskunde tussen droom en drama*. Kok Agora, Kampen.

Have, H.A.M.J. ten. (1995), *The anthropological tradition in the philosophy of medicine*, Theoretical Medicine 16, pp. 3-14.

Have, H.A.M.J. ten. (1997), *From synthesis and system to morals and procedure - the development of philosophy of medicine*. In: Carson, R.A. and Burns, C.R. (eds.): *The Philosophy of Medicine and Bioethics. Retrospective and Critical Appraisal*. Kluwer Academic Publishers, Dordrecht (in press), 1997.

Inlow, W.D. (1946), *Medicine: its nature and definition*, Bulletin of the History of Medicine 19, pp. 249-273.

Sandel, M.J. (1996), *Democracy's discontent. America in search of a public philosophy*, The Belknap Press of Harvard University Press, Cambridge (Mass.) and London.

Szumowski, W. (1937), *L'histoire de la médecine et la réforme des études médicales*, Librairie Gebethner & Wolff, Cracovie.

Szumowski, W. (1949), *La Philosophie de la Médecine, son histoire, son essence, sa dénomination et sa* définition. Archives Internationales d'Histoire des Sciences 2 (9), pp. 1097-1139.

Welie, J.V.M. (1994), *In the face of suffering. Prolegomena to a philosophical foundation of clinical ethics*, Universiteitsdrukkerij, Nijmegen

Welie, J.V.M. (1995), *Viktor Emil von Gebsattel on the doctor-patient relationship*, Theoretical Medicine 16, pp. 41 – 72.

BRUNO CADORÉ

A HERMENEUTICAL APPROACH TO CLINICAL BIOETHICS

One of the principles of clinical ethics can be defined as the ability to sustain the involvement of different parties in the critical reflection of possible action. It is necessary as first to have to consider the uncertainties, which are inherent in ethical reflection, before careful execution of decisions.

CLINICAL ETHICS AND CONTEXTUAL PHILOSOPHY

The method that I am going to put before you is not claimes to be the best method, but is one of many other valuations that we have made. One of its characteristics must be immediately underlined. The analysis has been used *after* the event has occured in situations involving difficult ethical problems. The methodological choice is as follows (B. Cadoré, 1995, 1995, 1997).

Frequently professional bodies are confronted with demanding ethical problems to which they have to make an immediate response, demanding extra time, stress and organisation. However, if they have some understanding of the considered ethical opinions, it greatly helps them to make better ethical decisions in the long run.

Advance analysis of ethical situations can provide an invaluable resource which can gradually be absorbed into the system, and from which professional bodies can draw, in the event of difficult ethical decisions which have to be made.

When decisions of this kind have to be made, it is imperative to keep the wellbeing and feeling of the person or group involved at the forefront of the analysis, and not, in any way, to judge them, but rather measure the impact of the final decision upon their future lives.

These ethical dimensions from analytical reflection, are to be integrated by the professional bodies, who are involved in caring for people facing such difficult ethical decisions.

As a result of this research, professionals bodies attending an ethical meeting, have been asked to present a hypothetical ethical problem, to be proposed and discussed, in order to elucidate their understanding.

Preparation asks for a short written proposal clarifying the lines of discussion and argument, to make the process easier.

This reflective work will have several successive steps. From this context we hope to draw a perception of the philosophy considering the clinical ethics in any given case.

This requires to select from two choices of method:

C. Viafora (Ed.), Clinical Bioethics. A Search for the Foundations, 53-59.

- We would like to define norms but the real aim of ethics is also to explicit the context of the difficult situation. Knowing this context it's easier to let emerge the ethical reflection of the different bodies. Thanks to this reflection the different bodies are able to understand what is part of the difficulties of the situation and what is part of the medical, social, cultural and personal context. This is what we call « to set a distance »; It's only then that we can judge the pertinence of the norms for this decision;
- It seems necessary to distance ourselves from a limited acceptance of ethical reflection which is not only a normative one. We have to put the philosophy in context, and to consider ethics in the global context. There should be a global dialogue between different parties to clarify and appreciate ethical problems.

There is then a choice when certain parties have to make a decision in biomedicine, for them to refer to ethical-decision making resources, if there is a need for this. They involve themselves into the situation according to the way they look over to their meaning of life.

Not only may they refer to their own point of view, but they are able to tap into the global ethical thinking. So medicine is now involved in the ability to enhance the health. But it is possible to discover a distance between these representations of health, life and the meaning of these representations that professional bodies would like to have (according to their own meaning of life).

Measuring the ethical impacts, arguing a decision, giving oneself the means to assume the responsibility of an action, these are the three main dimensions, of a clinical ethical approach, each of them being seen from a hermeneutical point of view.

THEORETICAL FRAMEWORK FOR INTERPRETATION

The theoretical reference to clinical ethics considered as contextual philosophy must be then be looked for on a hermeneutical point of view. The reference to Paul Ricoeur is important: this philosopher proposes to consider the sensible action as a text to which the main rules of the hermeneutical method would be applied (P. Ricoeur 1986). Coming after Max Weber who said that the aim of human sciences was « a behaviour orientated in a sensible way », Paul Ricoeur proposes to apply the criteria of textuality in the concept of sensible action in order to bring out the characteristic of legibility to set an interpretation of the action. The philosopher takes into account 4 criteria of textuality which define the passage from the speech to the text and on the basis of which, in an analogical manner, an interpretation of an action can be made. The speech is always given in the present time; when it is given as a text, this intentional exteriorisation shows the excess of the meaning, excess which is the echo of what has been said over what is said. What has to be understood in the speech goes beyond the materiality of what is formally said, excess which is shown not only in the event of speech but also in the echo it has in the listeners who, having listened to it, will echo it.The meaning is then always working.

Concerning the action, the first characteristic of legibility corresponding to this passage from the speech to the text is the « writing down of this action»: the sensible action becomes an object of science and of interpretation only it is an objectivation equivalent to the writing down of a speech. The meaning of the action comes then out of a simple event of the action showing the distance between that event and the internal logics of the action which have to be understood.

It is not only the act itself at a certain time of the history, which is important for that history, but really the logics which are in it and which are actualized in a way, as a nodal moment of achieving these logics.

This distinction is important in order to elucidate more what is set in action in the ethical making out of a taking a decision; Here again, putting the speech in relation to the text can help precise the conditions of understanding of an action. The speech refers to the person who speaks as we could think that the action refers to the persons who acts. However when a speech is translated in a text the reading of this speech let us perceive a distance between the verb meaning of the text and the mental intention of its writer. This dissociation constitutes the true issue of the written speech in a text.

For Paul Ricoeur, the same remark can be made about the action: he underlines the characteristic of autonomisation of it. An action comes apart from the one who acts and develops its own consequences, which give the action its true socialized dimension. Because of that, the action is exposed to interpretation. So for the ethical understanding, in order to know who is responsible we must consider how this action takes place in the socialized current matters. The implication of the responsibility is the one of the importance of the action in the orientation of the social time and its determinative structure. In other words what is concerned in the responsibility of the biomedical action is the impact of that action in the spreading of a certain relation to the social representations: here the representations of illness, health, and medical efficiency.

Then a third criteria of legibility of the action can be proposed. A speech tries to describe, express or represent a world. In the intersubjective situation of the dialogue in which a speech is delivered, this world can be either the one the speakers have in common, or the one which can be confronted to the world expressed by the speaker. But when the speech is given in a text, that one uses a meaning which is free from the only reality that the speech wanted to show.The world of the reader of the text, cut off from the situation, is then opened to other dimensions which were inherent in it. As Paul Ricoeur said, the references to a text open to new dimensions of our being in this world. It is the same for an action considered as a text. When the meaning of an action goes beyond the pertinence of an act in an actual situation, the context in which this act is done is revealed in an open meaning which is as a new mode of being part of the readers world. Analogically to references of a text, the context of an action, the decisions taken before it, the logics going through it, the roles of those building it are as many elements that allow the interpretation of an open meaning.

Then comes the fourth criteria of legibility of a sensible action considered as a text. Everyone can read the action proposed as a message, taken away from the only intersubjective relation and everyone would be able to -or have to- read the meaning of it, could consider his own involvement in the world as concerned by what has been left by the involvement of others. By what has been left, there could be a critical dialogue which could contribute to define the orientation of an act, laid down in common. Taking into account this point of view, Paul Ricoeur talks of a human action as an open work: everyone can -or must- be considered as a heir. At the same time this sets an act of interpretation as inherent in the involvement in the action, and this action as being in the line of responsibility which goes beyond the fact to have punctually to answer to successive acts which are laid down. In the biomedical field, It is then possible to consider that the succession of medical decisions involves not only the respect of criteria of good practices but also and specially the inscription of a certain meaning of the medical function: the parties from biomedicine discover themselves as having been set heirs of it, without deciding it.

FROM A THEORETICAL FRAMEWORK TO A CLINICAL ETHICS METHOD

When we become conscious that we are heirs of the meaning of the medical function, then a clinical ethical method can be settled as the intelligibility of a decision refers to the meaning of the function it expresses. That's why such a method tries to conceptualise the emergence and the promotion of an ethical subjectivity of the biomedical different parties. Considered as a text, the biomedical action could be the object of a « dialectic » approach between its explanation and its understanding. We can then elucidate the relation between the intentional dimensions and the justified ones, and at the same time we can make clear the relation between the involvement of the people and the institutional dimensions of the action. So putting the involvement of the parties in a context is really what is questionned here. Nevertheless it is not because we are talking about the notion of context that we aim to develop a « situationist » approach of ethics but on the contrary we would like to enlarge the ethical conscience of the different parties so that they become conscious that their involvement in the action has an ethical implication, as far as the quality of the medical cares and the orientation of the meaning of their function are concerned.

Such an hermeneutical approach of clinical Ethics is different from a method which consists in problem solving but it does not deny the importance of it. An hermeneutical approach of clinical ethics can be defined more as a research approach: from the interpretation of the action, the conditions of the ethical creativity of the involved parties must be determined. Such an approach is part of a certain definition of bioethics:

> "interdisciplinarian approach of a critical elucidation of the significance and the limits of the biomedical innovations in order to let the parties have an ethical creativity about the relation the individual and social human being has with illness and health ".

Two of the other complementary approaches to this research, considering the interpretation from an ethical point of view, must be underlined:

- a psychological approach: the emotional experience of the carer is taken into consideration,

- a sociological approach: the part of the institutional and social dimensions of cares in difficult situations is taken into account.

The practice in clinical ethics shows how the situations which are lived as difficult are really difficult. To the real clinical difficulty the different medical parties must take into account the psychological and social issues which do not directly depend on the biomedical action. Moreover we must consider the anthropological point of view: we must take into account the cultural dimension of the implementation of the medical function, specially of the therapeutical aspect. It is true that if we can consider health as a function of modern knowledges (how to put these knowledges into practices, as Michel Foucauld wrote) it is essential to consider this function as a mediator between Man and his wish to be healthy. It means that the reflection on the function of therapeutist is quite determinant for the ethical understanding of the clinic.

Considering all this, talking about the clinical ethics, we can talk of a process « research-action ». This will give us the means to have a critical reflection on the conditions of the ethical creativity of the different parties and this will lead us to take into account a new aspect of responsibility. This part of responsibility - and here the theoritical framework we have talked about is quite pertinent - can be described from the characteristics of legibility of a sensible action:

a- if we consider that taking a biomedical decision is a nodal moment of achieving logics included in the action, the responsibility then will be defined with respect to these logics which in fact determine the ethicity of the action;

b- since a sensible action enlarges the field of responsibility, we must value not only the decision itself but also the responsibility of the functions of medicine promoted by such an orientation of the action.

c- The biomedical action gives some representations which are interpreted. Among these interpretations, certain conceptions of the human living will be promoted and that is one of the object of responsibility.

d- So not only a certain quality of medicine is to be promoted but also the integration of medicine in the global field of promoting the ways to certify the human dignity and the service of it.

To which conditions is it possible to put forward the importance of dignity in each human being? How is practising medicine influenced by the social and cultural context in which some representations of this dignity are built up? To which kind of interpretations of this dignity is medicine handed over by a society? Can we say that medicine works out the promotion of human being, independant from these representations and interpretations?

So we can see that such an approach is apart from a simple normative approach and leads to a true criticism of normativities. Thanks to this approach it would be possible to understand the meaning of medicine that Emmanuel Lévinas describes as « la catégorie anthropologique du médical ». This open meaning is not only the one of the intersubjectivity in which the caring relation exists but it also shows firmly the cultural, social and political importance of medicine.

THE FUNCTION OF SETTING A DISTANCE

So in this clinical ethics approach, the aim is to work on the interpretation of the action so that we can make clear the points on which the responsibility of the parties is involved. We are not going to describe in detail this approach (method) but we would like to define how this method, while anlysing the situation, proposes to set such a distance that it leads to an hermeneutic of the responsible creativity.

a- narrating the situation is the first way of setting a distance.

We must take time to narrate a situation which has been lived as difficult, which has let us perplexed or unsastified. In doing so, we are given the means to make the different determinant elements of the situation be objective and at a distance to bring the subjective reactions to light. Setting the framework in which the ethical involvement of the actors has been asked, knowing that this involvement isn't the objective action.

b- studying the ethicity of the action is the second way of setting a distance.

When the action has been told, and when it is consider objective, it is possible to read it in different ways so that we can identify the elements which are part of its ethicity (that means how it makes the persons engage their own ethical capacity). Here are three important ways of reading the action because they'll help to set a distance.

1- The importance of the responsibility with regard to the caring relation itself;
The way we consider the body has an impact on the balance between the objectivation of the body and the subjectivation of the patient
The way we consider the suffering, how the care can help the person to write this illness in its own history
The way we consider health and the normative logics which influence it so that some possible alienations by normative pressures without any critics can be met;

2- The meeting of those constraints must be cleared up. In situations lived as difficult on an ethical point of view, it is clear that some different constraints exist: the constraints of the illness itself, those of the history of the person in his or her environment, the constraint of the care institution and its organisation, the one of social options, even the political ones in which the care is given. Facing those constraints the aim is not to deny them but, identifying them explicitly, to be able from those constraints to be creative.

3- Setting a distance needs also to take into account the different relations to time. We can suppose that if it often seems urgent to decide an action in a clinical practice it is often because we don't look at all the different relations to time of the different actors: the person, the caring team, the caring institution, the health policy. Knowing this interaction, the parties can carry on involving themselves in their creativity as without doubt it has an importance in the evolution of the situations.

c- A third way of setting a distance is, from that thinking back over situations, reflecting on the medical functions assumed by a team in a given pathological situation.The caring teams are very often quite impressed by intuitive « perceptions » of all those functions assumed through orientations given to their specialty. Those teams have very often too felt that even if they can do their best in their practice They have no power on the general orientations which are laid on them. Setting a distance with the taken decisions, a team can question whether these decisions correspond to the values they prize or not. They are then able to understand the meaning of their practice.

To look for the meaning in clinical ethics is not to define the meaning of the practice but to consider that the practitioners have to elucidate the meaning they would like for their practice. Carrying on a research promoting an hermeneutic of the creativity will prevent us from forgetting that the aim of the ethical creativity is to be involved into the practice, as we know that it is this practice which will be read by others. That's why, as Paul Ricoeur said, we need the analysis of an action as sensible action and action to be carried on or not.

Bruno Cadoré, Centre d'Ethique Médicale, Université Catholique, Lille, France.

REFERENCES

Ricoeur P. (1986), *Le modèle du texte: l'action sensée considérée comme un texte*, in *Du texte à l'action. Essais d'herméutique II,* Coll. Points- Essais, Paris, pp.205-235.

Cadoré B. (1995), *La methodologie en éthique clinique*, in: *Quelle éthique pratique en medicine?, Editions* E. C. C. E. , Bruxelles, pp. 17-43.

Cadoré B. (1995*), Ethique clinique et débat bioéthique*, in "Le Journal de l'infermière de neurologie et de *Neurochirurgie"*, 1-2.

Cadoré B.(1997*), L'éthique clinique comme philosophie contex*tuelle, Fides ed., Montreal.

c. A third way of setting a distance is, from that thinking back over situations reflecting on the medical function assumed by someone in a given method [illegible]

[illegible]

REFERENCES

[illegible]

MICHAEL PARKER

A DELIBERATIVE APPROACH TO CLINICAL BIOETHICS

A 'RIVALRY OF CARE' CASE

In their book, 'The Patient in the Family', Hilde and James Lindemann-Nelson describe the case of a man whose daughter is suffering from kidney failure[1]. She is spending six hours, three times a week on a dialysis machine and the effects of this are becoming increasingly hard for her and her family to bear. She has already had one kidney transplant, which her body rejected, and her doctors are unsure whether a second would work but are willing to try if they can find a suitable donor. After some tests the paediatrician privately tells the father that he is compatible and therefore a suitable donor.

It may seem inconceivable that a father would refuse to donate his kidney to his daughter under such circumstances. Yet he does refuse and justifies his decision both on the grounds that the success of the transplant is uncertain and also on the basis of his concerns about the implications of the operation itself for him and his family. He is frightened and worried about what would happen to him and his other children if his remaining kidney were to fail. But he is ashamed to feel this way and cannot bear to refuse openly so he asks the paediatrician to tell the family that he is in fact not compatible. Whilst having some sympathy, she says she cannot lie for him and, after a silence, the father says, 'OK then I'll do it. If they knew that I was compatible but wouldn't donate my kidney, it would wreck the family'.

But, why should this decision wreck the family, ask the Lindemann Nelsons? Does a father have a special obligation to donate his kidney to his daughter? What is it about families and the values that underpin them which leads to the expectation that parents will sacrifice themselves for their children (and in particular for the child who is ill)? What is it about modern patient-centred medicine that intensifies such expectations?

The case is used by the Lindemann Nelsons because they believe it suggests that there is a conflict in health care between two sets of values; those individualistic values which underlie patient- centred medicine and the communitarian values which sustain families and communities. They argue that modern medicine's overriding focus on the benefit of the individual patient has distorted the ways in which family members interact with one another and in particular with those who are sick. They argue that at times of stress families often adopt the individualistic values of the medical world and this leads them unintentionally to trample on the values and concerns that sustain families. It is with this tension, they suggest, that the father wrestles in the case described.

C. Viafora (Ed.), Clinical Bioethics. A Search for the Foundations, 61-71.

WHO AM I?

The claim that there are important tensions between the values of patient-centred medicine and those which sustain families and communities reflects an ongoing and important contemporary debate in bioethics (and in ethics more widely) between what have been called 'individualistic' approaches and those which have come to be known as 'communitarian'[2]. The conflict is one that is characterised by Michael Sandel and other communitarians as one between two conceptions of what it is to be a moral subject[3] which in turn presuppose two conceptions of the self.

The Communitarian Self

The Communitarian analysis of the case offered by the Lindemann Nelsons urges the father to seek a resolution of his moral problem in an answer to the question 'who am I?' where his identity is to be seen as informed by his membership of a community (in this case, a family) rather than through an analysis of rights[4] or a 'balancing' of principles'[5]. As Kukathas and Petit suggest,

> [For communitarians] the end of moral reasoning is not judgment but understanding and self-discovery. I ask, not "what should I be, what sort of life should I lead?" but "who am I?" To ask this question is to concern oneself first and foremost with the character of the community which constitutes one's identity[6].

Sandel too, argues that,

> I [should] ask, as I deliberate, not only what I really want but who I really am, and this last question takes me beyond attention to desires alone to reflect on my identity itself [7].

At the heart of this communitarian approach to the moral which urges us to emphasise the values which sustain families and communities over those of autonomy and patient choice, is the ontological claim that the moral world consists of fundamentally and essentially 'socially- embedded' beings who draw their identities, and their moral values, from their constitutive attachments to a 'community'.

Whilst communitarians would deny that their approach has the implication that individuals are less real than communities, their approach requires them to place more value on social values than 'individualistic' values. Moral reasoning for the communitarian as a result ought not to be characterised in terms of free rational choice and the pursuit of one's goals and conceptions of the good but should instead be seen as a search for an understanding of one's identity or role, an understanding which can only be achieved through a grasp of one's constitutive attachments to particular communities and values.

The communitarian critique of individualism is based at least partly upon the claim that there are aspects of our social embeddedness which it is impossible for us to escape; aspects which are constitutive of our identity as people. Communitarians go on to argue that the individualistic demand for individual rights and the associated attempt to escape our attachments, to escape who we really are is both unethical and unhealthy. Communitarians argue that the attempt to escape one's constitutive attachments and one's social identity is profoundly damaging not only for society but also for individuals themselves. And he draws our attention to the ills of contemporary liberal society as evidence for the truth of his argument.

In fact, argue the communitarians, however hard we try, it is in fact simply impossible for us to escape our constitutive attachments because they are aspects of our identity.

Thus it is that communitarianism has both a descriptive and a prescriptive element. The moral subject is said to be constituted by his or her embeddedness in a range of constitutive social relationships. And such relationships are of intrinsic value and ought to form the basis of moral judgments.

The Individual Self

Interestingly, Sandel argues that the deontological liberal too, whose approach it is which is rejected by the Lindemann Nelsons and other communitarians as 'individualistic', agrees that the question of who I am, or of the nature of the self is at the core of moral deliberation[8]. In contrast to the communitarian however, the individualist is said to conceive of the moral subject in terms of the autonomy and the free choice of the individual 'free chooser', rather than in terms of a being constituted by his or her embedded ness in a constellation of social and communal values. And this leads to an approach to ethics which emphasises the values of autonomy and patient choice over those of community and family.

The individualist argues that such freedom or autonomy is a necessary condition of the very possibility of the moral, and this means that autonomy ought to 'trump' other values[9]. As Sandel explains,

> For justice to be primary, certain things must be true of us. We must be creatures of a certain kind, related to human circumstance in a certain way. In particular, we must stand to our circumstance always at a certain distance, conditioned to be sure, but part of us always antecedent to any conditions. Only in this way can we view ourselves as subjects as well as objects of experience, as agents and not just instruments of the purposes we pursue[10].

The basis of an emphasis on autonomy is thus not the ends we choose but the capacity of us to choose them and such capacity depends upon the free and independent nature of the subject. As Kant argues, in response to the question of what makes the moral possible,

> It is nothing else than personality, i.e., the freedom and independence from the mechanism of nature regarded as a capacity of a being which is subject to special laws (pure practical laws given by its own reason)[11].

The emphasis on autonomy characteristic of individualistic approaches to ethics is therefore grounded in a conception of the subject or self as free and independent of exactly those features of the social world that the communitarians see as constitutive.

The power of the emphasis upon the needs and interests of individuals lies in its demand that any workable understanding of the moral world must relate in a meaningful way to the actual decisions which individual people have to make in their everyday lives. That is, it ties us to the moral concerns of real people and leads naturally to a conception of morality which is concerned with how individuals ought to live their lives in a world which is made up of individuals with competing conceptions of the good.

Individualists stress the differences between people, and essential separateness, rather than their similarities and shared interests and constitutive interdependence. They argue that it is because the human world is made up of individual people each with their own desires, interests and conceptions of the good, each possessing the ability to freely choose their own way of life, that moral concerns are a central feature of the lives we live. Our moral language reflects a shared need as individuals to work out ways of living together, again as individuals, and reflects also the difficulties we face in so doing.

That is, it presupposes an individualistic ontology. In this respect, individualists have thus sometimes been led to the conclusion that societies and communities are less real than the individuals which make them up and argue that communities ought not to have value over and above the value of the individuals of which they are constituted.

Thus it is that individualism too has been argued to comprise both a descriptive and a prescriptive aspect. Individualists claim not only that the world consists of individual people each of whom has the ability to freely choose their own way of life, but also that the needs and interests of individuals constitute the highest good. Hence what ought to be of most value ultimately is the actual freedom of individuals to make these kinds of choices. It is this that justifies the importance of upholding the rights of individuals against the demands of their communities. It is essential that any account of the morally complex world in which we live is at least sometimes able to do this. For, the emphasis upon the value of respect for the needs and interests of individual people undoubtedly resonates with one of our most important ethical intuitions.

A Conflict between models of the self

Both the individualist and the communitarian seek an explanation of the moral in an answer to the question of what it means to be a moral subject. Each rejects the other's answer to this question on the grounds that it is incapable of providing such an explanation. It is inevitable that any moral theory is going to have to have something to say about what it is to be a moral subject. I shall argue however, that both the individualist and the communitarian conceptions must be rejected in favour of an intersubjective and deliberative account and that this has important and far-reaching implications for the practice of bioethics, some of which I shall tease out in the final section.

THREE REASONS FOR REJECTING THE INDIVIDUALIST MORAL SUBJECT

It seems to me that the communitarian is right to reject the individualist model and the grounds for such a rejection can be grouped under three headings. I have explored these arguments more fully elsewhere and for reasons of space, I merely state them now[12].

The first of these grounds might best be collected under the heading, *The Impossibility of Moral Understanding* and draws together arguments from both philosophy and psychology which suggest that the individualist account of morality must be rejected because it is not possible to provide an explanation of the development of moral understanding from an individualistic epistemological perspective. For, the very possibility of moral understanding and moral language, it is claimed, is dependent upon the social dimension of human experience. Ludwig Wittgenstein's 'private language argument' is one powerful argument to this effect in which he argues that the very possibility of meaning and hence language depends upon the existence of standards of established social practice[13]. But this is not the only argument of this kind. Alasdair MacIntyre in After Virtue, for example, argues that,

> In so far as persons must be understood as partly individuated by their membership of traditions, the history of their lives will be embedded in the larger narrative of a historically and socially extended argument about the good life for human beings'[14]

The second group of arguments are those which claim, against the individualist, that the having of moral problems and moral identity at all depends on the fact that *we are 'socially embedded'*. That is, it is claimed, we are all inevitably located in social, intersubjective networks from which we draw our identity and that the liberal conception of the subject as divorced from such networks inevitably comes at a price. For, as Michael Sandel writes,

> To imagine a person incapable of constitutive attachments such as these is not to conceive an ideally free and rational agent, but to imagine a person wholly without character, without moral depth. [15]

Perhaps the strongest proponent of this type of argument is Charles Taylor who argues that to be a self at all is to be an essentially moral being located within what he calls evaluative frameworks and that such frameworks are inevitably linguistic and hence social.

> This is the sense in which one cannot be a self on one's own. I am a self only in relation to certain interlocutors: in one way in relation to those conversation partners who are essential to my achieving self-definition; in another in relation to those who are now crucial to my continuing grasp of languages of self-understanding- and, of course, these classes may overlap. A self exists only within what I call 'webs of interlocution'[16]

The third groups of arguments, (which I consider the weakest), are those which attempt to describe the *Unacceptable Social Consequences of Individualism*. Communitarians sometimes argue that historically the over-emphasis on rights in liberal democracies has had unacceptable consequences both for societies and individuals (i.e. the breakdown of traditional structures such as the family) and for this reason should be rejected[17].

Whilst I have my doubts about the strength of the third group of arguments in a world in which perhaps the most striking moral challenge is the oppression of individuals by communities, the combination of these arguments taken together means that communitarians are right to call for the rejection of what I have called elsewhere 'overly individualistic' approaches to ethics[18].

THREE REASONS FOR REJECTING THE COMMUNITARIAN 'EMBEDDED MORAL SUBJECT'

It seems to me however, that the communitarian argument for the 'socially embedded subject' must itself also be rejected for three sets of reasons which, again for reasons of space, I shall simply state here.

Firstly, the explanation of morality in terms of the 'socially embedded self and of 'constitutive attachments' means that communitarianism is incapable of recognising *the moral status of the individual.* Feminists for example have argued that whilst communitarianism is very good at describing the benefits of community, it says very little about the damage caused by families and communities and says nothing for those at the periphery of societies for whom we expect moral theory to have special concern. Taken to its logical conclusion, communitarianism seems capable of justifying the oppression of minorities and of the weak by the majority, of the novel by the traditional[19]. And whilst we might agree with the communitarians that overly individualistic approaches to ethics must be rejected we would surely not want to reject with it that which is valuable about the individualistic approaches; namely a recognition of the moral status of the individual. For this would be to throw out the baby with the bathwater.

Secondly and following from the above, the communitarian approach is, it is argued, incapable of providing an *explanation of social change* or of the need for the critical moral reflection, creativity and criticism necessary for the change and development of communities. Another way of saying this is to say that communitarianism is incapable of providing an account of how the individual can come to have an effect upon the society within which they live and upon their constitutive values and relationships[20 21].

Thirdly, Jürgen Habermas has argued that it is *not in fact possible to identify the shared values required by communitarians*[22]. The breakdown of shared values and traditions identified by communitarians brings into question the viability of the communitarian project itself. For, when we look around us there appear few if any candidates for the shared values upon which a communitarian New World might be built. We live in a world characterised by diversity in which candidates for the role of paradigmatic communities are revealed to be as often the sites of conflict and violence as of mutual support[23]; a world in which it is not possible to identify the kind of shared values or traditions upon which a communitarian morality might be founded[24].

A RESOLUTION? THE DELIBERATIVE MORAL SUBJECT

Both the individualist and communitarian models of ethics must be rejected. But where does this leave us? If we wish to elaborate a coherent moral theory[25] and, if appeal is no longer possible either to the kind of detached, individual, rational decision making called for by the liberal individualist or to communitarian shared values and traditions as the basis of ethical decision making in health care, how are we to approach the making of ethical decisions of the kind confronting the father at the beginning of this paper? What seems clear is that any coherent explanation of the moral will have to be one capable of capturing the insights of both communitarianism and individualism whilst avoiding their weaknesses and pitfalls and what this means is that it must be capable of capturing both the value of the individual voice and the moral status of the individual whilst at the same time of recognising the intersubjective and social context of morality and the value of social relationships and their various manifestations.

It is worth pausing here for a moment to reflect upon the interdependent nature of the relationship between the two sets of arguments I have identified for the rejection of individualism and of communitarianism. For it is an important feature of each of these arguments that such rejection is in each case put in terms of the necessity of the other to any coherent account of the moral. The argument that individualism must be rejected, for example, is based on the claim that recognition of the role of the social is a necessary element of any coherent explanation of morality. The argument for the rejection of overly social accounts on the other hand, is phrased in terms of the necessity of a recognition of the role of the individual.

My point in juxtaposing the arguments in this way is to suggest that both the social and the individual are together necessary and it is their combination that makes a coherent account of the moral possible. I want further to argue that these features of our moral world are jointly and together only explicable in terms of the actual relations between people in the intersubjective contexts which constitute their everyday lives with others. For it is only here, in the intersubjective relations between people, that the community meets the individual and vice versa. It is here that morality is elaborated and here that the maintenance and the transformation of social practice occur. This is to suggest following Harre[26] and Shotter[27] and other discursive psychologists that the primary social reality is neither the individual nor the community but people in conversation and that discourse is the developmental fundamental of human experience. To quote Alasdair MacIntyre from After Virtue,

> Conversation, understood widely enough, is the form of human transactions in general [28].

This must indeed be the case, I suggest, for the reasons above and because it is through such 'conversations' that we are introduced into the world of human affairs and negotiate our identity and our moral concerns. It is also here that we discover the ethical voice with which we reflect upon and change the nature of our relations to our community and other people. From this perspective it is possible to begin to recognise the particular value, and indeed the necessity, of the engagement of human beings in the negotiation of the meaning of their own lives and the nature of their relations with those around them, with those who constitute their communities.

Hence, within a moral framework of this kind is it possible to capture, as neither individualists nor communitarians are able, both the value of communal life and the moral significance of the individual ethical voice. It is to claim that it is neither the freedom of the abstracted individual nor the emphasis of community values which ought to be given a special place in the constellation of values but the deliberative relationship between the two. It is also to claim that the deliberative negotiation of meaning is the developmental fundamental of human experience and that it is this that makes the moral possible[29].

IMPLICATIONS FOR BIOETHICS

What then are the implications of this approach for bioethics? There are several key features of an approach such as this and I shall attempt to outline these very briefly in conclusion.

The value of 'making sense'

Firstly, to adopt this perspective is to argue, as I have already suggested, that the deliberative search for moral meaning is at the core of what it is to be human in a world with others. This is to locate morality and the search for moral meaning very firmly at the centre of human life. To adopt this perspective therefore is to recognise the particular value of the engagement of human beings in the attempt to 'make sense' of their lives and the nature of their relation with those around them. It is also by these means to recognise as neither individualists nor communitarians are able, *both* the value of communal life *and* the moral significance of the individual ethical voice. Whilst placing an emphasis on joint narrative and deliberation therefore, this approach nevertheless has the advantage of providing, as communitarianism does not, space for a critique of accepted or traditional values on the basis of a respect for the deliberative nature of human experience. For whilst respect of this kind is capable of capturing our social embeddedness it is *also* capable of recognising that individuals have a right both to be protected from, and to have a voice in, their community.

To assert the value of making sense and of the deliberative elaboration of the self s in many respects to follow Alasdair MacIntyre who argues for a conception of the moral life as one constituted by engagement in a conversation with history and tradition and other people in an attempt to establish the narrative unity of one's life. It is also to align oneself with Charles Taylor's claim that the identity of the self is inextricably linked to its sense of the significance and meaning of the situations it encounters in life and this is to see, as does Ronald Dworkin, life as a series of 'challenges' which must be addressed[30].

Subsidiarity and participation

Secondly, it follows from the emphasis on the value of 'making sense' that ethical decisions are best made and in fact might only be capable of being made by those most closely involved, for whom such choices are meaningful or significant, and this is to suggest that the process of making ethical decisions ought to adhere to a principle of 'subsidiarity'. Nevertheless, such an approach is also and perhaps primarily one which emphasises the participation of all those who have a legitimate interest and this means that the requirement that decisions be made by those most likely to be affected needs to be balanced against a responsibility to ensure that all who have a legitimate interest have a voice. This is to suggest that decision making in bioethics will need to take a range of different forms, from the establishing of public consensus conferences about ethical issues of widespread public or even global concern, to conversations between doctors, patients and families or within families themselves about the ethical questions raised by a particular case or treatment option and in some cases, perhaps even most, this will means that decisions will be made by the patient alone, or in collaboration with his or her doctor. The balance between subsidiarity and participation in any particular case being decided and hence justified, deliberatively.

This is an argument for the democratisation and decentralisation of ethics. For, whilst the philosophical analysis of ethical problems and ethical theory and the elaboration of biomedical principles can be useful in creating a framework for the discussion of ethical problems, the resolution of such problems in an ethical way involves the creation and maintenance of ethical fora of the kind I have described in which those who have a legitimate interest in a case can engage jointly in the process of making moral sense of the situation. This is to argue for a genuinely participatory, democratic and deliberative bioethics and such a perspective has profound and radical political implications both for the medical profession and beyond. A deliberative approach to ethics is grounded in an intersubjective model of what it means to be a moral subject in which moral decisions are reached be deliberation and justification in a social setting appropriate to this purpose.

The need for the elaboration of further deliberative principles

I have attempted in these final paragraphs to sketch out some of the features of an approach to the resolution of ethical question which is both intersubjective and deliberative. A fuller account of this approach would have to say more about the appropriate constitution and membership of a deliberative ethics forum and it would also have to say a great deal more about the modus operandi of such fora. If such a forum is to be both genuinely deliberative, intersubjective and ethical it will have to follow procedural rules designed to maximise participation and deliberation and genuine debate. It will not be immune for example from the power dynamics of the world outside and work will have to be done to develop a deliberative model which is able to go some way at least towards the minimisation of factors such as this which impede both the free flow of deliberation and its orientation towards the resolution of dilemmas in the making of practical decisions[31].

THE 'RIVALRY OF CARE' CASE REVISITED

What does the adoption of the approach I have been attempting to sketch mean for the case with which this paper began?

Briefly, the first implication of an intersubjective approach is that it rules out interpretations of the case which suggest that it forces us to choose between individualistic values and those which might be said to be communitarian. Moral problems can only be understood, and so resolved, in intersubjective terms. This raises difficult hermeneutic questions about meaning. Taken from one side this will mean that emphasis will need to fall on what this decision means for the father and how it makes sense in terms of the way in which he understands himself, as a father and so on. In this sense and in the light of my earlier comments about subsidiarity, the decision about whether or not to donate his kidney should be a decision for the father, either alone if he so wishes (or with support if he needs it). Nevertheless, the case raises difficult ethical questions for the clinician too. For the clinician has an important role both in the decision to donate and also, importantly, in the earlier decision whether or not to be tested for compatibility. One has to ask why the father pursued testing if he had no intention of donating his kidney (if this is the case). Was he adequately informed? Had he access to counselling before and during the decision?

Whilst the final decision in this particular case must rest with the father, the case also raises ethical issues which can only be resolved by a broader deliberative process. What ought to be the hospital's policy on informing and counselling patients prior to, during and after, testing? How ought potential donors be informed? What degree of confidentiality can they expect? These decisions will need to be made in a forum involving perhaps clinicians, counsellors, patients and so on. Such a case might also throw up broader ethical questions which can only be resolved politically or at the national level. An example might be the development of an ethical national policy on transplantation.

Finally, in the case as presented, there seems no reason for the clinician to inform the family of the father's results. She might say simply that there are no kidney's available for transplantation, for example. But there is no reason for her to give more information than this. It may be that because of inadequate counselling (or other reasons) the father has let the family know that he is being tested and has put himself in the position of having to lie to them. This raises once again the ethical issues relating to what counts as adequate counselling. Assuming this is not in fact the case, the father's confidentiality must be respected. Hence his fear that the family will be torn apart by this knowledge might be avoided.

Michael Parker, Oxford Institute for Ethics and Communication in Health Care Practice, Lecturer on Medical Ethics, University of Oxford, United Kingdom.

NOTES

[1] Lindemann N., H and J. (1995), *The Patient in the Family*, Routledge, New York, p.1.

[2] Parker M. (1999), *Ethics and Community in the Health Care Professions*, London, Routledge.

[3] Sandel M.(1982), *Liberalism and the Limits of Justice*, Cambridge, Cambridge University Press.

[4] Lindemann Nelson, H. and J. Ibid. , pp.1-30.

[5] Beauchamp T. and Childress, J.(1994), *Principles of Biomedical Ethics* (fourth edition), Oxford, Oxford University Press.

[6] Kukathas, C., and Petit, P., Rawls (1990), *A Theory of Justice and its Critics,* Cambridge, Polity, p.106.

[7] Sandel, M., (1982), ibid., p.180.

[8] Sandel, M., ibid.

[9] Dworkin, R. (1977), *Taking Rights Seriously*. London.

[10] Sandel, M. (1982), ibid. p.11.

[11] Kant, I. *Critique of Practical Reason*. Trans byBeck.L.W., Indianapolis: Bobbs-Merill, 1956[1788].

[12] Parker, M.(1995), *The Growth of Understanding*, Aldershot, Avebury.

[13] Wittgenstein, L. (1974), *Philosophical Investigations.*2nd edition, Oxford, Blackwell, n.150-200. See also: Taylor, C. (1989), *Sources of the Self,* Cambridge, Cambridge University Press; Vygotsky, L.S.(1978), *Mind in Society,* Harvard. Mead, G.H.,(1934), *Mind, Self and Society*, Chicago; Donaldson, M. (1978), *Children's Minds*, London, Fontana, MacIntyre, A.(1981), *After Vi*rtue, London, Duckworth, p.197.

[14] MacIntyre A.(1981), *After Virtue*. London, Duckworth, p 197.

[15] Sandel, M. ibid, p.179.

[16] Taylor, C. (1989) *Sources of the Self.* Cambridge, Cambridge University Press, p.36.

[17] See for example many of the works by Amitai Etzioni such as 'The Spirit of Community' New York, Crown, 1993.

[18] Parker M. (1995), *The growth of Understanding*, Aldershot, Avebury.

[19] Parker, M. (1996), *Communitarianism and Its Problems.* Cogito, November.

[20] Parker M. (1995), *The Growth of Understanding*. Aldershot, Avebury.

[21] Mendus, S. Strangers and Brothers. *Liberalism, Socialism and the concept of autonomy*, in Milligan, D. and Watts-Miller W.(1992), *Liberalism, Citizenship and Autonomy*, Aldershot, Avebury, p.13.

[22] As indeed has Alasdair MacIntyre. See for example, After Virtue, ibid.

[23] Campbell B. (1995), *The London Independent*, March 16.

[24] Habermas, J. (1993), *Justification and Application: Remarks on Discourse Ethics*, Oxford, Polity.

[25] I would like to thank Matti Hayry for helping me to see that this is what I meant.

[26] Harre, R. and Gillett, G.(1994), *The Deliberative Mind*, London, Sage.

[27] Shotter J. (1993), *Conversational Realities*, London, Sage.

[28] MacIntyre A. (1981), *After Virtue*, London, Duckworth, p.197.

[29] Parker M. (1995), *The Growth of Understanding*. Aldershot, Avebury.

[30] Dworkin R. (1988), *The Foundations of Liberal Equality* in *G.* Petersen (ed.) The Tanner Lectures on Human Values.

[31] Habermas, J. (1993), *Justification and Application: Remarks on Discourse Ethics*, Oxford, Polity.

Lindemann Nelson, H. and J. Lindemann Nelson ... pp.1-30.

Beauchamp, T. and J. Childress, J.F. 1994. *Principles of Biomedical Ethics* (fourth edition). Oxford: Oxford University Press.

[illegible]

PART 2

TOWARD CLINICAL BIOETHICS INTEGRATING "INTERNAL MORALITY" AND "EXTERNAL MORALITY"

HENK TEN HAVE

"A HELPING AND CARING PROFESSION": MEDICINE AS A NORMATIVE PRACTICE

INTRODUCTION

The development of modern bioethics can be related to the complex process of interaction between internal and external morality in the area of healthcare. Evaluating in retrospect the birth of bioethics shows that there is a dialectic process: traditional ethics almost exclusively emphasized the internal morality of medicine, modern bioethics in contradistinction stressed the significance of external morality, whereas the present-day methodological criticism of bioethics articulates the unique interplay of internal and external morality. This dialectic process is related to different views of medicine as well as particular goals of medicine. In the traditional view, medicine was clearly a profession orientated towards diseased individuals; the focus was on prolonging life and curing disease. Modern bioethics encouraged a view of medicine as an enterprise or even trade or business, exchanging technological support and expert knowledge with the demands and needs of autonomous persons. Present-day critique reformulates medicine as a profession, but not focused on its own interests but more orientated towards communal needs, not merely answering to the particular wishes of individuals but rather focusing on what is necessary for care and support to guarantee citizens adequate functioning in society.

FROM TRADITIONAL MEDICAL ETHICS TO BIOETHICS

Traditionally, "medical ethics" referred to the deontology of the medical profession, to codes of conduct which consist partly of ordinary moral rules, partly of rules of etiquette, and partly of rules of professional conduct (Downie, 1974). In this sense bioethics has the following characteristics:

a) It is essentially a set of problems that focuses on the internal morality of medicine, *viz.* those values, norms, and rules intrinsic to the actual practice of health care. Medicine is not considered a merely technical enterprise that can be morally evaluated from some exogenous standpoint. On the contrary, the professional practice of medicine always presumes and implies a moral perspective or point of view; therefore, what is judged to be sound medical practice is determined by the shared rules and standard procedures of the practice.

b) It is related to the professionalization of medicine. Through this historical process emphasis is placed on the common good, and this was combined with an appeal to the self-interest of the members of the profession. Social recognition could only be gained on the basis of a strong internal organization and self-imposed

C. Viafora (Ed.), Clinical Bioethics. A Search for the Foundations, 75-97.

standards of behavior. Self-regulation by physicians and a special style of life, structured in terms of high ideals, duties, and virtues, could promote the physician's image, and thus the power and prestige of each member of the medical profession.

c) Since it was primarily concerned with explicating norms and formulating standards of professional conduct, medical ethics and etiquette have been segregated for a long time from general intellectual history (Fox, 1979). Moreover, before the 1960s, medical ethics was not a subject frequently discussed in public fora and the extant literature. Apparently, there was a consensus of opinion concerning the moral commitments of those who provided medical care, and the explication or codification of these commitments was regarded as the principal concern of medical professionals.

Since the 1960s, professional medical ethics has gradually detached from its traditional deontology (ten Have and van der Arend, 1985), although there are important phase differences between the North-American and European countries; especially in some East and South European countries the emphasis in medical ethics still is on "medical deontology". In North-western European countries professional medical ethics more and more is subsumed under "health care ethics", or "bioethics". These new terms tend to indicate that the discipline of ethics not only includes problems that arise in the physician-patient relationship, but also a number of moral problems posed by other health care professionals, as well as moral issues created by the health care system, and public policy issues engendered by biomedical advances and the results of research. This is also illustrated by the dramatic increase in the number of publications on moral problems in medicine and health care authored by non-physicians. Consequently, the range of problems that properly is subsumed under the rubric "medical ethics" is considerably enlarged; there are ever new and more complex moral issues, and new participants emerge to participate in an intensified set of medico-ethical debates.

Thus, the result of the gradual transformation of medical ethics is two-fold.

First, it has produced a new professional - the health care ethicist or "bioethicist" who possesses a specific body of knowledge and particular cognitive skills; who publishes in specialized journals, participates in newly-formed societies, and teaches in newly-established centers, institutes, and departments.

Second, it has produced a new socio-cultural interest in medico-moral matters of significant public concerns - particularly in those countries where advanced biomedical technology permeates public as well as private life. "Bioethics" is a way of publicly addressing, explicating and debating problems generated by science and technology.

The outcome of the above-mentioned transformation process is more salient in the U.S.A. than in most European countries. Moral problems in health care are generally approached in U.S. bioethics in a specific way, - more analytic and applied as is usually the case in many Continental approaches.

The aim of the first part of this contribution is to analyze the dominant conception of bioethics which seems to prevail in bioethical debates anywhere in the world. This conception originated in the American rather than in the European context. However, interest in alternatives conceptions and methods of bioethics is currently increasing. Criticizing the dominant conception, therefore, at the same time leads to the question whether there are specific characteristics of European approaches to moral problems in health care. Is it possible to identify typically European approaches in the area of bioethics? The second part of this contribution developes significant ideas in European medical ethics.

EVALUATING THE CURRENT STATUS OF ETHICS

Today there is growing concern that the results of the transformation from traditional medical ethics to modern bioethics are unsatisfactory. Professionalization and institutionalization of ethics received an enormous stimulus because both the adequacy and the relevance of medicine's internal morality were put into question. Professional ethicists have placed more and more emphasis on the crucial role of external morality: the principles, norms, and rules operative in society that bear on medicine and are frequently codified in law. Thus, for some, medicine and health care are nothing more than interesting "intellectual" phenomena with respect to which general ethical theories, principles, and rules may be applied.

This shift from internal professional to external morality and the predominant interpretation of medical ethics as "applied ethics" encouraged physicians to criticize present-day medical ethics for its lack of attention to the practical vicissitudes of health care, for its theoretical biases, and its conceptual alienation from clinical reality (Editorial, 1990; Vandenbroucke, 1990; Fulford, 1994; Welie, 1998; Wulff, 1998).

It is also claimed that the conceptual ground of medical ethics is too limited and even reductive when seen from the perspective of the tradition of philosophical ethics itself. Must medical ethics be conceptualized as applied theory rather than as reflective practice? (Baier, 1985; Kass, 1990; Murray, 1994; Zwart, 1995; Evans, 1998).

In addition, it has been suggested that there is a serious discrepancy between the public's attention to moral questions and the actual impact of ethical analysis on the routine practices of medicine, as well as the current direction of medicine's development. Moral issues tend to appear every day, but how successful are "bioethicists" in addressing these novel issues? To be sure, the media reflect a constant fascination with the myriad of moral problems in health care, but what concrete effect do these debates have on physicians' decisions in daily clinical medicine, on nursing practice, and on public health policies? Arguably, such "discrepancies" result from the very conception of medical ethics in our time (ten Have and Kimsma, 1990).

'APPLICATIO' AND 'PRINCIPIUM'

During the last thirty years, a popular and unique view of medical ethics as a new discipline separated from philosophy, theology as well as medicine has emerged. The growing appeal of this new discipline among public and scientific circles of opinion leaders can be attributed to the empowering combination of two traditional notions from the history of moral philosophy: 'application' and 'principle'.

The current conception of medical ethics reflected in the mainstream of scholarly literature is that of *applied ethics*. In Beauchamp and Childress' well-known textbook, biomedical ethics is defined as

> the application of general ethical theories, principles and rules to problems of therapeutic practice, health care delivery, and medical and biological research (Beauchamp and Childress, 1983, p IX-X).

Instead of the theoretical abstractions of traditional moral philosophy, applied ethics can contribute to analyse dilemmas, resolve complex cases and clarify practical problems arising in the healthcare setting. The practical usefulness of applied ethics not only manifests itself in biomedicine, but it has a wider scope. In the *Encyclopedia of Applied Ethics*, the following definition is presented:

> Applied ethics is a general field of study that includes all systematic efforts to understand and to resolve moral problems that arise in some domains of practical life, as with medicine, journalism, or business, or in connection with some general issue of social concern, such as employement, equity or capital punishment (Winkler, 1998, p. 192)

A distinction is made between three major areas of applied ethics: biomedical ethics, business and professional ethics, and environmental ethics. However, the table of contents of the four encyclopedia volumes show a wide range of topics covered, such as archaeological ethics, censorship, divorce, electronic surveillance, gun control, nuclear power, vegetarianism, and wildlife conservation. Applied ethics can extend to almost any area of life where ethical issues arise. 'Application' here has a double connotation: it indicates that ethics is available for what we usually do, it applies to our daily problems; but it also is helpful, practical, in the sense that ethics is something to do, it works to resolve our problems.

The second characteristic of the dominant conception of medical ethics is the focus on *principles*. If ethics is conceived as applied ethics, then subsequent reflection is needed on what is being applied. The emerging consensus that principles should provide the answer to this query, is coherent with the moralities of obligation that have dominated modern ethical discourse, especially since Kant. Behavior in accord with moral obligations is considered morally right. The morality of behavior is a morality of duty. Morality is understood as a system of precepts or rules people are obliged to follow. Particularly in the early days of bioethics, when medical power was strongly criticized, and the rights of patients were vehemently emphasized as requiring respect, the moralities of obligation presented themselves as a common set of normative principles and rules that we are obliged to follow in practice. As Gracia (1999) pointed out, the *Belmont Report* in 1978 was influential because it was the first official body to identify three basic ethical principles: autonomy, beneficence and justice. A basic principle was defined as a general judgment serving as a basic justification for particular prescriptions and evaluations of human actions. From these principles, ethical guidelines can be derived that could be applied to the biomedical area. About the same time, Beauchamp and Childress, in the first edition of their book, introduced the four-principles approach, adding 'nonmaleficence' to the above three principles. In their view, principles are normative generalizations that guide actions. However, as general guides they leave considerable room for judgment in specific cases. Various types of rules are needed to specify the principles into precise action guides.

Although Beauchamp and Childress have considerably nuanced their theorical framework in later editions, their work has contributed to the conception of medical ethics that is currently dominanting the practical context, in ethics committees, clinical case-discussions, ethics courses, and compendia and syllabi. This conception is sometimes called "principlism": the focus is on the use of moral principles to address ethical issues and to resolve conflicts at the bedside (DuBose, Hamel and O'Connell, 1994). Belief in the power of principlism is sometimes proselytizing. Raanan Gillon, for example, argues that the advantage of the four principles not only is that they are defensible from a variety of theoretical moral perspectives, but also that "they can help us bring more order, consistency, and understanding to our medico-moral judgments" (Gillon, 1986, p. viii). Later, Gillon used the principles-approach to develop a major scholarly project, the voluminous textbook *Principles of Health Care Ethics* (Gillon, 1994). Over 100 authors discuss in 90 chapters all possible ethical dilemmas in modern health care, employing the analytical framework of the four principles. In his Preface, Gillon confesses that he is inclined to believe that the four principles approach can encompass all moral issues, not only those arising in health care. Principlism apparently is a universal tool; it provides a method to resolve all moral issues in all areas of daily life, whatever the personal philosophies, politics, religions, moral theories of the persons involved.

The dominant conception of medical ethics, arising from the combination of the notions 'application' and 'principle', implies the following set of interdependent presuppositions (ten Have, 1990b):

a) medical ethics is application of ethical theory and ethical principles;

b) there is a body of available ethical theories and principles, and rules to be applied to a variety of practical, biomedical problems, in particular the framework of the four principles;

c) professional ethicists have a special expertise in applying ethical theories and principles, whereas non-ethicists (e.g., physicians) merely provide moral problems for applied ethics;

d) medical ethics is general ethics applied to medicine. That is, the context in which these problems arise is not unique in the sense of being characterized by specific values which generate special problems. Indeed, the medical context is viewed as a practice-ground for a new profession of biomedical ethicists; and

e) the aim of medical ethics is to proffer practical recommendations and prescriptions based on or deduced from ethical theories and principles.

This set of presuppositions to some degree clarifies why medical ethics is perceived by many as an independent discipline. For example, there is the view that ethics should perform four tasks: conceptual clarification, analysing and structuring arguments, weighing alternatives, and advising a preferable course of action (de Beaufort and Dupuis, 1988, p 19-20).

The central contribution of medical ethics is therefore restricted, but nonetheless powerful. It does not necessarily result in judgments regarding what we should do. The ethicist provides the topography of arguments, and objectifies the options. The ethicist regards himself as a disinterested and neutral observer of medical practice, who is in the best position to weigh moral alternatives.

THE DOMINANT CONCEPTION: CRITIQUE

Only recently have the presuppositions underlying the prevailing conception of applied ethics been critically questioned. Consider the following three arguments:

(a) In daily medical practice, medical ethics focuses on *mid-level principles* - respect for autonomy, beneficence, nonmaleficence, and justice. These principles are applied to dilemmas, cases, and problems encountered in the practice of health care. From a specific principle, guidelines or recommendations can be derived in order to resolve various problematic situations. Yet there is no single rational criterion on the basis of which to decide which principle is overriding; there is no definitive scheme for ordering principles and for choosing between them. As long as the principles of applied medical ethics are not integrated into some broader theoretical framework they tend to lead to conflicting judgments about which actions and social policies one ought to carry out. Even if one proceeds from some articulated moral theory (e.g., consequentialism, contractarianism) one cannot evade the chaos of conflicting moral judgments (B. Brody, 1988). The lack of agreement on which moral theory to apply on concrete medical cases could make applied ethics counterproductive. Confronting physicians and medical students with a variety of conflicting but plausible theories, applied medical ethics may be seen to give no moral guidance but to reinforce the belief that whatever is done in problematic situations, some moral theory will condone it, another will condemn it (Baier, 1985). Thus the primacy of applied ethics and the deductivist model of applying general moral theories and intermediate principles can only lead to an inadequate way of conceiving the relation of ethics to medicine (Jonsen, 1990).

Because the dominant conception of medical ethics focuses on the application of principles, norms and rules, it is rather loosely embedded in philosophy, and lacks a more encompassing critical, theoretical perspective on its own practical activities.

(b) Secondly, the dominant conception has developed within a *particular cultural context*. The fundamental ethos of applied medical ethics, its analytical framework, methodology, and language, its concerns and emphases, and its very institutionalization have been shaped by beliefs, values, and modes of thinking grounded in specific social and cultural traditions. Nowadays, the medical ethics literature serves as one of the most powerful means by which to express and articulate these traditions. However, the medical ethics literature only rarely attends to or reflects upon the socio-cultural value system within and through which it operates. Scholars usually assume that its principles, theories, and moral views are transcultural. H.T. Engelhardt (1986), for example, distinguishes between *two* levels: that of secularized pluralistic society and that of the many particular moral communities with competing visions of the good life. Bioethics, in his opinion, should focus on the societal level, speaking across gulfs of moral discourse; it is a common neutral language, a secular moral grammar, guaranteeing a peaceable society. The most interesting task of ethics is on the first, societal level: promoting and defending, in the context of health care, the general secular moral language of mutual respect. Critics agree that this is an important task; but it flows from a rather thin or minimalist conception of ethics (Callahan, 1981). Ethics is conceptualized as procedural; it is the regulation of social relations through peaceable negotation. In order to speak the language of mutual respect, all other moral languages must be pacified.

But why should we abstain from our particular moral language in favor of a neutral common language? This question points to an important problem: how neutral is the common neutral language? Is Engelhardt's language itself not the specific moral language of a specific moral community? Is this language itself not the expression of a commitment to a certain "hypergood" (Taylor, 1989), in particular, the good of universal and equal respect and self-determining freedom, - primal values in the liberal tradition? Such questions assume that the value of mutual respect and rights to privacy are not decontextualized standards but themselves expressions of community-bound agreements.

Only recently has there been an increasing awareness that a critical examination of the socio-cultural context is necessary if we are to better understand the strengths and weaknesses of this currently dominant conception of medical ethics. The medical sociologist R. Fox (1989), for example, has shown how the political norms of liberalism and individualism are very much characteristic of North American bioethics. By stressing the autonomy and rights of individuals, other significant considerations (e.g., community and the common good, duties and responsibilities) have been neglected, as have critical philosophical questions concerning the value of medical progress and personal and public health in communal life. Although interest in the philosophy of medicine in Europe in general seems to emphasize the social aspects of medicine and the common good, rather than individual autonomy, the dominating conception of medical ethics in particular countries such as the Netherlands seems in many respects not significantly different from that in the U.S. where liberalism and personal autonomy are stressed.

(c) Another criticism of the dominant conception of medical ethics is its *inattention to the particularities of the practical setting*. Moral theories and principles are necessarily abstract and therefore not immediately relevant to the particular circumstances of actual cases, the concrete reality of clinical work, and the specific responsibilities of health care professionals. By appealing to principles, norms or rules, applied ethics may fail to realize the importance of concretely lived experience of health care professionals, as well as patients. The moral agent is taken to have an abstract existence. This point is critically elaborated by contemporary philosophers. Ethics, according to B. Williams (1988), does not respect the concrete moral subject with his personal integrity. It requires that the subject gives up his personal point of view and exchanges it for a universal and impartial point of view. This is, Williams argues, an absurd requirement, because the moral subject is requested to give up what is constitutive for his or her personal identity and integrity. The idea that knowledge of normative theories and principles can be applied to medical practice simply ignores the fact that moral concerns tend to emerge from experiences in medical settings themselves. A similar issue is raised by Ch. Taylor in his *Sources of the Self*, in which morality and identity are considered two sides of the same coin (Taylor, 1989). To know who we are is to know to which moral sources we should appeal. The community, the particular social group to which we belong, is usually at the center of our moral experience. Even the use of ethical language depends on a shared form of life. The Wittgensteinian notion that our understanding of language is a matter of picking up practices and being inducted into a particular form of life is germane here.
In short, medical ethicists should become more appreciative of the actual experiences of practitioners and more attentive to the context in which physicians, nurses, patients, and others experience their moral lives, e.g., the roles they play, the relationships in which they participate, the expectations they have, and the values they cherish (Zaner,

1988). The physician-patient relationship is neither a-historical, a-cultural nor an abstract rational notion; persons are always persons-in-relation, are always members of communities, are immersed in a tradition, and participants in a particular culture.

From these points of view two conclusions may be drawn (ten Have and Kimsma, 1990):

(1) Morality is something we all participate in; medical ethics in particular is not the result of esoteric knowledge; anyone involved in the medical setting is ipso facto a moral participant and "expert" at least with regard to moral experience and intuitive knowledge.

(2) The moral experience inherent in health care practices must be taken into account -- more than the conformity of these practices with pre-existing ethical theories. From the perspective of applied medical ethics, abstracting from the reality of practices and appealing to moral principles and rules outside these practices, are necessary conditions to criticize health care practices. The problem, however, is not only how such a standpoint external to concrete practices is possible, but also whether appeals to external morality are not vain without intimate knowledge of the morality internal to the practices in question (Jensen, 1989).

ALTERNATIVES

Given the criticisms noted above, an alternative approach to and conception of medical ethics is clearly needed -- a conception that provides a more comprehensive understanding of the nature, scope, method, and application of ethics in the contemporary health care context. It will be necessary to re-connect medical ethics with both a general philosophical standpoint and the concrete practice of medicine.

In order to achieve a more adequate understanding of the possibilities for such re-connection, it may prove useful to outline promising new perspectives.

a) In response to the theoretical and methodological weaknesses of applied ethics, *new approaches* to medical ethics are available: phenomenological ethics (Zaner, 1988), hermeneutic ethics (Carson, 1990; Leder, 1994), narrative ethics (H. Brody, 1987; Hunter, 1988; Newton, 1995), and care ethics (Tronto, 1993). Furthermore, traditional conceptions have been revitalized, notably the new casuistry (drawing from the classical casuistic mode of moral reasoning) (Jonsen and Toulmin, 1988), and the virtue approach, emphasizing qualities of character in both individuals and communities (Drane, 1988; Pellegrino and Thomasma, 1993).

b) Other approaches have emerged due to the recent appreciation of the relevance of the *social and cultural matrix* in which medical ethics necessarily operates. For example, D. Callahan (1990) has argued that the ethical problems generated by the need for health care resource allocation and for the formation of new health policy have forced us to explore the goals and ideals of medicine as well as the meaning of health in modern society. However, thoughtful empirical research into the value systems relevant to the formation of moral issues in health care is relatively rare. For example, data acquired from sociological value research as well as the methodological approaches of social scientists (e.g., Halman, *et al.*, 1987; Inglehart, 1990) are virtually unknown and therefore ignored in medical ethics. The rigid distinction between descriptive and normative ethics could, in part, account for the absence of empirical value studies in medical ethics. Only recently, however, are there signs that a more positive interaction between medical ethics and the social sciences can be achieved (Weisz, 1990). What is particularly striking is the interest in so-called empirical medical ethics. The focus of medical ethical research is shifting from applying ready-made ethics toward studying ethics-in-action (Arnold and Forrow, 1993). A variety of research methods is used: participatory observation, questionnaires and interviews, decision analysis, quality assessment, preference polls. The common denominator is that qualitative and quantitative data are collected via the empirical study of ethical questions. Many of these studies are fascinating since they show the underlying value pattern of specific practices and the intrinsic norms which are operative in clinical work, for example in surgery (Bosk, 1979), genetic counseling (Bosk, 1992), intensive care (Zussman, 1992), neonatal care (Anspach, 1993), and nephrology (Lelie, 1999). Although empirical research in ethics can provide new and useful insights, and can be regarded as complementary to philosophical approaches (Hope, 1999), it is also troubled with fundamental problems (ten Have and Lelie, 1998). One of the basic questions concerns the moral relevancy of empirical data. Empirical research can help to explain and understand the attitudes, reasonings and motivations of the various actors in the health care setting, but empirical data in themselves can not justify how the actors ought to behave or what kind of decisions are morally justified (Pellegrino, 1995).

(c) A relatively new conception of medical ethics is so-called *clinical ethics*. It has emerged in response to the criticism that applied ethics is too far removed from the realities of medical practice. Clinical ethics aims to reorient medical ethics within the daily health care setting (Jonsen, Siegler and Winslade, 1986).

The extent to which clinical ethics differs from the prevailing conception of applied ethics can be characterized as follows:

(1) Interdependence of technical and normative dimensions of medical judgment. This interdependence which is at the basis of clinical ethics, is repeatedly underlined by recent work in philosophy of medicine. It is argued that clinical medicine is intrinsically a moral enterprise since it presumes a healing relationship between physician and patient. Since value judgments are pervasive in clinical decisions, moral concerns are inseparable from certain technical concerns, e.g., the correct diagnosis and the most effective treatment.

(2) Insider perspective.

The realities of clinical decision-making are crucial for the identification of ethical issues such that the ethical problems that arise in the practice of surgery are not identical to those that arise in pediatrics, obstetrics, or gynaecology. Moreover, they are not of the same nature "medically", since they differ with respect to risks and benefits. Specifically, the insider's perspective allows for the determination of whether risks, in routine investigations, are low, or whether they are substantial with questionable benefits. Thus it is asserted that an insider perspective is not only required to direct attention to the ethical questions that arise in clinical encounters, but properly to acquire empirical data relating to the process and outcome of these clinical encounters: How do patients and physicians actually make decisions? What moral options are involved? What are the effects of personal and professional values in reaching clinical decisions?

(3) Method of induction.
Instead of utilizing a deductive method by which general theories and principles are applied to practical moral dilemmas, an inductive methodology should be utilized which begins with a careful analysis of specific empirical conditions. This view, in part, accounts for the renewed interest in classical casuistry (Jonsen and Toulmin, 1988). The casuistical method includes the search for paradigm cases in which a particular moral maxim for right action is clearly applicable. Analogies are then proposed regarding cases in which, due to different circumstances, other moral maxims appears less suitable. The casuist thus explores a range of cases and scenarios forming more to less plausible arguments. Thus the factual circumstances of a case are extremely relevant: by modifying them they reveal new insights. The casuist's task is to determine the degree to which relevant moral maxims "fit" the particular circumstances. Even more: the casuist seeks to determine which factors, personal preferences, and social conditions and values are relevant enough to be judged as significant "moral facts".

(4) Clinical ethics is an inherent function of medicine itself.
This is a logical consequence of the points just mentioned. When physicians consider ethics as intrinsic to their craft, then the ethical analyses of medical decisions cannot proceed from an externally imposed system; essentially, they are an inherent, second-order function of clinical medicine itself.

From this survey of criticism and alternative approaches it is concluded that medical ethics is presently dominated by a limited conception of ethics -- the *application* of moral theories and principles to cases. This conception depreciates the fundamental internal morality of the professional practice of medicine by stressing external morality. This conception also reveals a lack of interest in the empirical realities of clinical medicine and neglects the socio-cultural value-contexts in which medical care is provided. In short, a broader framework for a practicable medical ethics is needed.

EUROPEAN APPROACHES: TRANSCENDING PRAGMATISM

It is prima facie problematic to identify typically European approaches to medical ethics. It has been argued earlier (ten Have, 1995), that if there is a difference in the medical ethics literature between Europe and North America, it is that European authors put more emphasis on: 1) the historical perspective of ethical issues, 2) the sociocultural context, and 3) substantive normative viewpoints.

This different focus leads to a somewhat different agenda of bioethics. In addition to concentrating on attempts to analyse and resolve practical problems, European literature shows a desire to overcome pragmatism by raising philosophical questions concerning the human condition, the perfectibility of the human being, the impact of biotechnology, basic concepts such as health, disease, and disability, and the epistemology of medical science. The feeling is that the dominant conception of medical ethics is loosely embedded in philosophy, thereby lacking a more encompassing critical, theoretical perspective on its own practical activities. The success of this conception flows from its applicability to practical problems, its educational value and its pragmatic concentration on elucidating and resolving dilemmas and problems. Doing so, medical ethics itself has been transformed into a more or less technical approach, technethics. This is a paradoxical result. Moral issues arise from an almost exclusively technological orientation to the world and a predominant scientific conceptualization of human life; we try to address these issues with a conception of ethics, itself impregnated with scientific-technical rationality. The dominant conception of medical ethics still seems very much under the spell of the Marxist formula that philosophy should change the world, not interpret it. Unfortunately, through its emphasis on pragmatism and applicability, it cannot change the world of medical science and technology, since it is too much part of it.

The paradox is exemplified by the role of the human body in bioethical discourse. Viewed from the tradition of philosophy of medicine, bioethics emerges from the anthropological criticisms of medicine, as argued above. Bioethics therefore originates from the recognition that medicine separates the individual person into subject and object, and that the human subject needs to be re-introduced into medical discourse. The best way to focus attention to the patient as a whole person and as an agent being in control of his or her own life, is to stress the autonomy of the individual subject and to demand moral respect for this autonomy. However, the emphasis on individual autonomy tends to neglect the significance of the human body. In most ethical discourse, there is no recognition of the special experiences of embodiment; it seems as if the autonomous subject is not embodied. Its body is merely the instrument through which the subject is interacting with the world. The subject is in full control of its body. It is imperative that the integrity of the body should be respected, as it is the prime vehicle of the autonomous person.

The moral principle of respect for autonomy in health care ethics apparently is associated with a popular image of the body as property (ten Have and Welie, 1998). When the individual person is regarded as autonomous subject, then the body is his private property. And the person is the sovereign authority with property rights over his or her body. Since autonomous individuals own their bodies, they have exclusive possession of it and they alone have it at their disposal. This concept of body ownership is increasingly important now in debates concerning transplantation, research, genetics and reproductive technologies. Property language in health care ethics is used to designate the locus of decision-making authority: the individual as owner is in control over his body. In view of the increasing medical possibilities to invade the human body as well as the potential of body parts for research and commercial purposes, it is necessary to protect the individual person against harmful and paternalistic interventions with and into his body. At the same time, the concept of body ownership is morally problematic. The distinction between person and body is contrary to the existential identity with our bodies and the self-experience of ourselves as embodied selves. In making such a distinction between autonomous subject, c.q. owner and a body, c.q. private property, bioethics seems to proceed from the same dualism which was criticized in the anthropological tradition. More so: it is apparently using a dualistic distinction between person and body, subject and object, - a distinction which has led to the emergence of bioethics itself. Whereas medicine tends to neglect the subject, bioethics tends to neglect the body (ten Have, 1998a; Zwart, 1998).

What has been one of the prime motivational sources for the growth of bioethical debate (a reductionist view of human beings as bodies without relevant subject) apparently is copied in bioethical discourse itself (a counterpart reductionist view of human beings as subjects without relevant body). Although the precise vocabulary has changed, in biomedicine as well as bioethics, a similar dualism of object and subject in regard to human beings can be recognized.

CONNECTING INTERNAL AND EXTERNAL MORALITY

The different emphases in European literature seem to have a common denominator: they focus on the dialectic connection between the internal and external morality of medicine, without reducing one set of norms and values to another. It is heuristically assumed that on the one hand there are specific values, norms and rules intrinsic to the actual practice of medical care (the "internal morality"), on the other hand, values, norms and rules prevailing in social, cultural and religious traditions that function as external determinants of medicine (the "external morality"). The dominant conception of medical ethics proceeds from a too strong distinction between these two sets of values, norms and rules, as well as an over-estimation of the relevancy and importance of the external morality. In order to obtain a better understanding of the interaction of both moralities, it is necessary to establish a theoretical framework relevant to medical practice in order adequately to take account of the norms and values inherent in the practice of medicine, but it requires at the same time sufficient detachment in order to provide a critical normative perspective on medical practice.

The problem is how this task can be accomplished. How to develop a theoretical perspective on medical ethical issues that connects philosophical reflection with the everyday realities of medical practice? Such perspective not only aims at elucidating specific bioethical problems, but it intends critically to examine various conceptions of bioethics that purport to deal with such problems. It should also make clear why and how such bioethical problems appear, reappear, and even disappear in medical discourse; why certain problems emerge in various health care practices and others do not; and how such problems can be discussed and even resolved during daily interactions between physicians, nurses, patients, hospital administrators, and others.
In programmatic attempts and debates pursuing these objectives, at least four steps can be distinguished.

(1) The first step is to examine the methods of clinical ethics in different health care contexts in order to obtain a better understanding of the *internal morality* of these practices. This will require to formally undertake both empirical research and philosophical investigations. A new theoretical perspective on bioethics can be developed only if we take seriously certain fundamental notions of clinical ethics (ten Have, 1990a). One of these notions is that there are internal standards and norms that govern professional medical practice. These internal norms are made salient by analyzing various health care practices. Recently, a revival of the concept 'internal morality' is noticeable. Especially changes in the health care system as well as financing structures have stimulated anew reflections on the nature of medical practice (Brody and Miller, 1998).

John Ladd introduced the concept to refer to norms governing medical practice; they determine what is good clinical medicine:

> Internal morality ... comprises moral norms relating to the clinical situations that depend on 'medical' considerations, such as diagnosis, prognosis, treatment plans, concepts of disease, and so on (Ladd, 1983, p. 212)

He contrasted it with the notion of 'external morality'; this notion includes the moral considerations that come from outside medicine; they are based on non-medical facts like social conditions, personal habits, and demands of individual patients and their families. The focus on the internal morality reiterates the view that medicine is a profession (as discussed earlier in this chapter; see also: Ladd, 1989). In this view medicine is not a morally neutral body of knowledge and technique; its moral content cannot be derived from the general morality of society. A full account of the content of the internal morality of medicine, as Brody and Miller (1998) acknowledge, requires further development of two constituents: the moral goals of medicine and the morally acceptable means for achieving those goals. The clinical practice of medicine is directed on a set of particular goals, a coherent range of good healing actions. As Brody and Miller point out these goals should not be too narrowly identified (interpreting 'healing' as 'curing a disease'); at the same time, even a comprehensive list of goals is limiting medical activities and requiring particular moral values rather than others. Medical practice also requires internal standards of appropriate performance. Promotion of a particular goal alone is not sufficient; it should go with morally acceptable means. Brody and Miller suggest four standards, originating in the nature of medical practice:

> 1. The physician must employ technical competence in practice...
> 2. The physician must honestly portray medical knowledge and skill to the patient and to the general public, and avoid any sort of fraud or misrepresentation...
> 3. The physician must avoid harming the patient in any way that is out of proportion to expected benefit, and must seek to minimize the indignity and the invasion of privacy involved in medical examination and procedures...
> 4. The physician must maintain fidelity to the interests of the individual patient... (Brody and Miller, 1998, p. 388).

These examples from the literature indicate the direction to go for future developments. The first step toward a reconnection of practice-internal and practice-external moralities is carefully to examine daily health care practices. Surely clinical ethics requires such a reorientation; yet for many practitioners "clinical ethics" does not suggest an alternative view of medical ethics, but only serves to further the application of ethical rules and principles to cases. That is, "clinical ethics" simply means "doing ethics in the clinical setting". Clinical ethics under this construal is simply a special case of applied ethics. The disadvantages of this approach can be overcome, however, if one introduces ethical discourse directly into the clinic, thereby retaining the prevailing values.

Clinical ethics, under this view, involves a new approach to ethics that is relevant to all health care practices. Clinical ethics becomes a radically different interpretation of ethics because it takes place within the clinical setting. It is possible, therefore, to profit from clinical ethics without reducing clinical ethics to applied ethics.

(2) The second step is to analyze and interpret the *external morality* governing health care practices. Making use of the results of recent social research and specific empirical investigations, this step requires the study of values, norms, and attitudes concerning medical-ethical issues. To date, value studies have only occasionally examined (patients') values regarding health, disease, dysfunction, disability, dying, illness prevention, and health care. These values in society need to be explored in order to understand more fully the value context in which current bioethical debates occur.

An example of this approach would be a research project that focuses on values regarding health, disease, dying, illness prevention and health care that are explicit or implicit in public policy documents concerning care for the chronically ill. What norms and values are reflected in public policy documents as well as actual public policy decisions for the chronically ill? Important values in this context are, for example, solidarity and justice. In the Dutch health care system, solidarity seems to imply not only that the community will take care of the ill and helpless but also that the weak will limit their claims to care when there is no longer any prospect of a meaningful life for them. The value of justice is significant as well since we seem to lack a guiding vision of how a just and good society should accommodate the special needs of its chronically ill members. In view of the growing prevalence of chronic illnesses, traditional concepts of solidarity and justice will become more problematic. The question will be how much society wants to afford to care for the chronically ill. But the issue is also how chronic suffering is valued in a particular society. In order to have a better understanding of the current bioethical problems in chronic health care practices, the normative context of such practices (in social debate, in public policy decisions, and policy documents) should therefore be further explored.

(3) The third step is the creation of *new theoretical perspectives* on health care practices. History of medicine as well as philosophy of medicine share a growing interest in the empirical realities of medicine. The so-called empirical shift in philosophy of science, some decades ago, has led to new approaches, e.g., several kinds of social constructivism (ten Have and Spicker, 1990). From this social constructivist perspective important contributions to medical theory have been made (Latour, 1987). From this point of view, diagnoses, diseases, medical knowledge, health care institutions are considered social constructions, which can be understood only in their *empirical* social and cultural context.

Ethics, philosophy and history of medicine may thus find common ground in creating new theoretical perspectives on health care practices. In any practice a complex set of activities guided by shared rules, cognition, action and normativity are inextricably linked. Focusing on the notion of *practice* as the common theoretical starting-point, the interdependence of the disciplines as well as the specificity of their expertise will become apparent. Philosophy of medicine analyzes the cognitive components of health care practices: concepts, methods, and ideas. Medical ethics examines the activities and action-guides embodied in health care practices as well as the values embedded in such practices. History of medicine studies the diachronical and synchronical construction and transformation of practices.

A critical evaluation of theories of medical practice is therefore necessary. The work of the Danish philosopher, Uffe Juul Jensen (1987), gives a useful and interesting example of a philosophical theory of medical practice. Jensen's theory is a conceptual framework as well as a heuristic instrument to study the problems of modern health care - such as those arising in the care of chronic patients - from moral, philosophical and historical perspectives. The modern health care system is a complex network of practices based on different historical traditions, embodying different values and using different methods. Jensen specifically distinguishes three kinds of practice-orientations that are woven together in the modern health care system: the disease-orientated practice, the situation-orientated practice, and the community-orientated practice. Obviously, a critical analysis of Jensen's specific viewpoints is necessary; nonetheless, the focus of his model for the interrelationship of knowing, acting and valuing in health care practices seem to be a promising starting-point for analysing and elucidating present-day moral problems in present-day health care.

(4) The fourth step is to develop a *new conception of bioethics* that illuminates and clarifies the complex interactions between the internal and external morality of health care practices. As a particular domain of philosophy, ethics proceeds from empirical knowledge, *viz.* moral experience. The moral dimension of the world is first and foremost experienced. Moral experience is humanity's way of understanding itself in moral terms (van Tongeren, 1988). Ethics is therefore the interpretation and explanation of this primordial understanding. Before acting morally we must already know, at least to some extent, what is morally desirable or right. Otherwise, we would not recognize what is appealing in a moral sense. On the other hand, what we recognize in our experience is typically unclear and in need of further elucidation and interpretation.

In short, we approach the moral dimension of the world from a set of prior understandings; they form the basis of our interest in what at first seems odd and strange to us, requiring us to continuously reconstruct the moral meaning of our lives. Such an interpretive perspective will be helpful for integrating the experiences disclosed in the clinical-ethical studies, as well as utilizing the insights gained from describing the value-contexts of health care practices.

INTERPRETIVE BIOETHICS

Overlooking the theatre of competitive approaches, one of the challenges to contemporary medical ethicists is to formulate a new conception and practice of medical ethics that can bridge the gap between the internal and external morality of medicine, as well as between medical empiricism and ethical normativism. It requires the development of a theoretical framework relevant to medical practice so that it may adequately take account of the norms and values inherent in the practice of medicine, but with sufficient critical distance so that it may provide a normative perspective on these practices.

Graber and Thomasma (1989) developed the unitary theory of clinical ethics out of a concern with the problematic relationship between theory and practice in medical ethics. Having examined various models of theory-practice relation (for example, the model of applied ethics), they believe the new theory will avoid the weaknesses of these models and combine their strengths. The Unified Clinical Ethics Theory (UCET) therefore can incorporate elements of the virtue, deontological, and consequentialist theories of ethics. The theory is summarized as follows:

> Certain conditions (C) are present in this case such that the probablity (x) exists that Value (V) A will be judged more important than B by (I) interpreters because the Principle (P) p' will be more likely to apply to the case than p" (Graber and Thomasma, 1989, p.194)

However, in this summary statement it is not obvious that it does indeed combine such theories. It is emphasizing the context of a case, the weighing of relevant values, and the role of interpreters but the normative justification for judging value A more important than B is that principle p' takes precedence over p".

At the same time, Graber and Thomasma consider UCET as a practical model of bioethical hermeneutics that combines both theory and practice. The hermeneutic aspect is repeatedly mentioned by the authors: all cases require interpretation; interpreters are involved in profound ways in analysing the case and balancing its important features (O.c., p.196); an essential part of making moral judgments is interpreting the fit between situation and principles (O.c., p.201).

However, the pragmatic orientation of UCET has possibly prevented a further elaboration of this interpretive point of view, so that it is unclear how radical the hermeneutic perspective really is: is it methodological hermeneutics, paying adequate attention to the interpretive components of medical practice, or is it hermeneutic philosophy, trying to develop a theory of interpretation and to explain medicine as a hermeneutic science? If the last focus prevails, the crucial question for ethics as a practical enterprise is not so much to clarify action guides and make moral quandaries controllable but rather to make them communicable.

Graber and Thomasma have not further developed their hermeneutic philosophy; it may seem that hermeneutics has simply been incorporated as a tool into a hybridization of virtue, deontological and consequentialist theories. Even the name "unitary theory" suggests an harmonious combination of different approaches, whereas in fact the authors are aiming at an encompassing ethical super-theory, absorbing specific ethical theories within a radical hermeneutic framework.

However, what really is innovative in their approach is the emphasis on the role of interpretation. The term "hermeneutic" can be misleading. It may suggest that now a particular school of philosophy is applied in the context of health care whereas the philosophical point of view is the emphasis on interpretation.

Hermeneutics

Originally, hermeneutics refers to the art of interpreting and the science of interpretation. As such, it was used in theology, law and philosophy, all concerned with interpreting the meaning of texts. It has come into prominence in the last century as a methodology characterizing the humanities and social sciences. Philosophers such as Schleiermacher and Dilthey have shown that not only texts but all human products need interpretation, and that hermeneutics involves not only the interpretandum but also the interpreter. Finally in our century, through the works of Heidegger, Gadamer and Ricoeur, hermeneutics evolved into a philosophy of understanding and explaining human existence.

Medicine as hermeneutics

Entering a new stage of the long-standing debate on the status of medicine, it has been argued that medicine has to be considered as a hermeneutical enterprise apparently presuming that medicine is not or not merely a natural science (Daniel, 1986; Leder, 1988; Svenaeus, 1999). The modern emphasis on information and empirical data has contributed to new understanding of diagnosis and treatment as the physician's interpretation of what concerns the patient and what can be done to help the patient. And metaphorically, the patient is conceived as a text that may be considered on different interpretive levels. It is important to reflect upon the typical preconditions of interpretation in medicine. The patient is usually understood through an anatomico-physiological model. The patient's body is made `readable' by the use of technology. The biomedical language of diagnosis and treatment reduces the overwhelming amount of information presented by the patient so that the standard medical case report reflects not the story of the patient's life but of the physician's relationship with the patient's illness (Poirier and Brauner, 1988).

It is also important to look at the effects of medical interpretation upon the interpreter. Interpretation seems to bring understanding and empathy. Interpreting symptoms involves understanding what is actually wrong with a patient and appreciating what he or she is going through. Interpreting the patient's illness arouses therefore an "affiliative feeling" in the physician-interpreter (Zaner, 1988).

Bioethics as interpretation

Some contemporary philosophers have argued that ethics is best considered to be a hermeneutical discipline. Ethics therefore can be defined as the hermeneutics of moral experience. Complex bioethical problems must be understood within the broader framework of an interpretive philosophical theory. Such a theory should concentrate upon four characteristic parameters.

a) Experience

The starting-point of medical activity is the moral experience of the patient. Through his illness he is confronted, in Zaner's words (1988, p. 65), with tears in the fabric of daily life. He presents himself to the physician as both puzzling and meaningful. The patient's symptoms are deeply textured by his biographical situation, with his beliefs, values, habits and life-style. To ascertain what is wrong requires an interpretation, the more so since there is an initial distance between patient and physician. The meaning of the individual human being who is the patient requires interpretation for two reasons: a) intrinsic strangeness; the experience of illness in this particular patient is unique and unusual; b) theoretical pre-understandings: the context in which the physician interprets the symptoms (e.g., the pathological models) is different from the context in which the interpretandum came into existence. It can reasonably be expected that moral experiences differ according to the interpretive models used in various health care practices and according to the specific complaints, illnesses and disabilities of the patients encountered in different health care settings. Different practices should therefore be examined and compared. Experiences are part of the text of life. But we never know whether our understanding of this text is adequate (van Tongeren, 1994).

b) Attitudes and emotions

For ethics, the fundamental question is not so much "What to do?" but rather "How to live?". It is *praxis* not *poiesis* that is important (van Tongeren, 1988). The moral relevancy of our actions should not be reduced to their effects; it is also determined by an evaluation of what we do in executing our actions. For example: the problem of experimenting with human embryos should not be settled by reference to future results, but should also raise the question: Why are we interested in scientific research? This change of focus implies a re-orientation from activity to passivity, from acts to attitudes and emotions. Moral experience involves primarily feelings, for instance, of indignation, confusion or contentment; secondarily, these emotional responses can be made the object of moral thinking (Callahan, 1988).

A sharing of moral experiences of patients and physicians, and of the emotions and attitudes involved, is therefore required for elucidating the relevant ethical issues of the case or situation. Understanding and defining the morally relevant facts of a case do not involve the identification of relevant general principles and the deduction of a set of rules from which the correct response to the problem can be derived. The role of medical ethics is not so much to explicate and apply ethical theories and principles but to interpret and evoke what is implied in moral experience. The notion 'applied ethics' suggests wrongly that we already know which moral principles and rules to apply. However, rules and principles are in fact answers to what is evoked or appealed to in a particular case. First of all, we need to understand what the moral experience of vulnerability and appeal to assistance really mean in this case. We need to discover why particular principles will motivate us in this case; why is there a particular ideal, rule or obligation? It requires close scrutiny of the medical situation in all its complexity.

c) Community

The interpretive reading of a patient's situation is not an individual doctor's affair. The medical prior understandings that orientate the interpretation are the sediments of traditional cultural assumptions concerning the nature of the world and the body, and the results of a specific historical evolution of medical knowledge. Interpretation presupposes a universe of understanding. This is a consequence of the so-called hermeneutic circle; in order to interpret a text's meaning, the interpreter must be familiar with the vocabulary and grammar of the text and have some idea of what the text might mean (Daniel, 1986). For man as a social being, understanding is always a community phenomenon: understanding in communication with others. The continuous effort to reach consensus through a dialogue with patients, colleagues and other health professionals, induces us to discover the particularities of our own prior understanding, and through that, to attain a more general level of understanding. This seems to reflect the experience of hospital ethics committees: analyzing a case in terms of moral principles leads to a stalemate but interpreting the moral experience of the concrete participants involved in this particular case usually leads to a consensus. Since the interpretation of moral experience takes place within the context of particular social practices, intimate knowledge of the historical, medical and scientific components of those practices is essential to the task of moral criticism. Ethics can not be practised without a high degree of engagement in medical work.

d) Ambiguity

Ethics primarily aims at interpreting and understanding moral experience. But moral experience is complex and versatile. It implies that every interpretation is tentative; it opens up a possible perspective. Definitive and comprehensive interpretation is non-existent. An interpretive approach always has an ambiguous status: more than one meaning is admitted. As Zaner (1988, p. 272) points out: "Every life is linguistically inexhaustible, there is always a richer tale to be told that can never be wholly captured in words, no matter how evocative they might be". That means that moral judgments and decisions which must be framed on the basis of understanding the thematic moral ordering of a person's life are fundamentally uncertain.

Implications

Interpretive bioethics will have important consequences for the competence and role of medical ethicists. Since the interpretation of moral experience takes place within the context of particular social practices, intimate knowledge of the historical, medical and scientific components of those practices is essential to the task of moral criticism. Ethics can not be practised without a high degree of engagement in medical work. Bedside medical ethics does not imply that only doctors can be ethicists. To comprehend the human terms in which actual moral dilemmas are experienced, the facts of medical practice must not be left to the doctors. But being able to do that requires a sharing of moral experiences of patients and doctors.

To implement the interpretive conception of medical ethics, more empirical study of actual decision-making processes is necessary. Like the empirical turn in philosophy of science, the current interest in empirical medical ethics is combining medical sociology and anthropology, history of medicine and philosophy to construct a more sophisticated view of moral experience in medicine.

Interpretive bioethics also has consequences for ethics teaching. If it is assumed that ethics is part and parcel of the routine practice of medicine, and that ethics is not an abstract exercise of moral reasoning but characterized by the emotion, complexity and ambiguity commonly involved in real cases, then lessons should be drawn for the method and goals of medical ethics teaching (ten Have and Essed 1989). Preference should be given to team-teaching in the clinical period using the format of patient conferences and case review, generally accepted in routine clinical work. The objective of this problem-orientated ethics teaching is first and foremost to increase the students' sensitivity to moral problems in everyday medicine.

Finally, interpretive bioethics will require a new rapprochement between ethics and philosophical anthropology (ten Have, 1998b). As described elsewhere (Ten Have, 1990c), during this century there has existed an undercurrent of philosophical criticism of modern medicine with very different manifestations: originally epistemological, then anthropological, now ethical. Particularly in health care, normative positions and moral theories are intimately connected with images of the human being. In the medical setting we cannot escape the question: what kind of human being do we want to realize in medical activities, what kind of person do we wish to respect, heal, inform, comfort in health care? However, the relationship between ethics and anthropology is a dialectical one. Instead of claiming the primacy of anthropology, with ethics based on anthropological theory and images of the human being, philosophy of medicine should also focus upon explicating the morality underlying anthropology itself. Ethical discourse can exemplify a particular image of man, but at the same time anthropological discourse itself is presupposing particular moral views. From the point of view of interpretive bioethics, morality is not something we choose, but a fundamental predicament we are already involved in before we even start to reflect upon it; such predicament, however, at the same time is an anthropological characterization of what is essential in human beings. Morality is choosing us, because we are primarily social beings. Ethical views are articulated and explicated because we are in a moral relationship with other human beings appealing to us.

The dialectical interaction of anthropology and ethics is helpful in regaining a view of man as social being, and therefore restoring the idea of moral community. Currently, we can notice an interest in communitarian approaches to bioethics (Kuczewski, 1997). They emphasize that cultural context and community are constitutive of the values and goals of individuals. Communal relatedness falsifies the idea of the unencumbered self, the idea of self-ownership assuming that the individual as an entity exists prior to the ends which are affirmed by it. Without societal culture our potential for self-determination will remain empty.

The face of the other makes us moral beings whether we like it or not, whether we choose to act accordingly or not. Morality is a social affair. Its inter-personal character makes it possible to scrutinize and criticize individual moral choices.

CONCLUSION

The data, insights, and theoretical notions obtained and analyzed in the previous steps require integration through developing a theory of medical practice with emphasis on its ethical dimensions, that can illuminate and clarify the complex interaction between the internal and external morality of various health care practices. Criticism of the dominant conception of applied ethics and principlism can help to articulate what kind of theory is needed and which conception of bioethics is most adequate to understand medical practice.

Henk Ten Have, Professor of Medical Ethics, Department of Medical Ethics, Philosophy and History of Medicine, University Medical Center, Nijmegen, The Netherlands.

REFERENCES

Anspach, R.R. (1993), *Deciding who lives. Fateful choices in the intensive-care nursery,* Univerity of California Press, Berkeley.

Arnold, R.M. and Forrow, (1993), *Empirical research in medical ethics: an introduction,*Theoretical Medicine 14, pp.195-196.

Baier, A. (1985), *Postures of the mind. Essays on mind and morals.* Methuen, London.

Beauchamp, T.L. and Childress J. F. (1983), *Principles of biomedical ethics* (2nd ed.), Oxford University Press, New York/Oxford.

Beaufort, I.D. de and DupuisH.M.(1988), *Handboek gezondheidsethiek,* Van Gorcum, Assen/Maastricht.

Bosk, C.L. (1988), *Forgive and remember. Managing medical failure.* The University of Chicago Press, Chicago/London.

Bosk, C.L. (1992), *All God's mistakes. Genetic counseling in a pediatric hospital.* The University of Chicago Press, Chicago/London.

Brody, B. (ed.) (1988), *Moral theory and moral judgments in medical ethics.* Kluwer Academic Publ., Dordrecht.

Brody, H. (1987), *Stories of sickness.* Yale University Press, New Haven and London.

Brody, H. and Miller, F.G. (1998), *The internal morality of medicine: Explication and application to* managed *care. Journal of Medicine and Philosophy* 23(4), pp.384-410.

Callahan, D. (1981), *Minimalist ethics. On the pacification of morality,* In: Caplan, A.L. and Callahan, D. (eds.), *Ethics in hard times.* Plenum Press, New York, p. 261-281.

Callahan, D. (1990), *What kind of life: The limits of medical progress.* Simon and Schuster, New York.

Callahan, S. (1988), *The role of emotion in ethical decision-making. Hastings Center Report* 18: 9-14.

Carson, R.A. (1990), *Interpretive bioethics: The way of discernment. Theoretical Medicine* 11: 51-9.

Daniel, S.L. (1986), *The patient as text: A model of clinical hermeneutics. Theoretical Medicine* 7: 195-210.

Downie, R.S (1974), *Roles and values. An introduction to social ethics.* Methuen, London.

Drane, J.F (1988), *Becoming a good doctor: The place of virtue and character in medical ethics.* Sheed and Ward, Kansas City.

DuBose, E.R., Hamel, R. and O'Connell, L.J. (eds.): 1994, *A matter of principles? Ferment in U.S. bioethics.* Trinity Press International, Valley Forge (Pennsylvania).

Editorial: 1990, Medical ethics: should medicine turns the other cheek? *The Lancet,* oct.6: 846-7.

Engelhardt, H.T. (1986), *The foundations of bioethics.* Oxford University Press, New York/Oxford.

Evans, M. (ed.): 1998, *Critical reflection on medical ethics.* JAI Press, Stamford (Conn.).

Fox, D.M. (1979), *The segregation of medical ethics: a problem in modern intellectual history. Journal of Medicine and Philosophy* 4: 81-97.

Fox, R.C. (1989), *The sociology of medicine. A participant observer's view.* Prentice-Hall, Englewood Cliffs, N.J.

Fulford, K.W.M. (1994), *Not more medical ethics.* In: Fulford, K.W.M., Gillett, G.R. and Soskice, J.M. (eds.): *Medicine and moral reasoning.* Cambridge University Press, Cambridge, U.K., p. 193-205.

Gillon, R. (1986), *Philosophical medical ethics.* John Wiley & Sons, Chichester.

Gillon, R. (ed.): 1994, *Principles of health care ethics.* John Wiley & Sons, Chichester.

Graber, G.C. and Thomasma, D.C. (1989), *Theory and practice in medical ethics.* Cotinuum, New York.

Gracia, D. (1999), *History of medical ethics.* In: ten Have, H.A.M.J. and Gordijn, B. (eds.): *Bioethics in a European perspective.* Chapter 1.

Have, H. ten and Arend,A. van der (1985), *Philosophy of medicine in the Netherlands. Theoretical Medicine* 6: 1-42

Have, H.A.M.J.ten and Essed, G. (1989), *An experimental case-conference programme for obstetric and* gynaecolgy *clinical students. Journal of Medical Ethics* 15: 94-98.

Have, H. ten and Kimsma,G.K. (1990), *Changing conceptions of medical ethics.* In: Jensen, U.J. and Mooney, G. (eds.), *Changing values in medical and health care decision making.* John Wiley & Sons, Chichester, p. 33-51.

Have, H. Ten, (1990a), *Ethiek tussen alliantie en dissidentie*. Inaugural lecture, University of Limburg, Maastricht.

Have, H. Ten (1990b), *Ein Hippocratische erfenis; ethiek in de medische praktijk*. De Tijdstroom, Lochem.

Have, H. Ten (1990c), *Verleden en toekomst van medische filosofie. Scripta Medico-philosophica*, Schrift , p. 5-18.

Have, H. ten and Spicker, S.F.: 1990, Introduction. In: Have, H. ten, Kimsma, G., and Spicker, S.F. (eds.): The *growth of medical knowledge*. Kluwer Acad. Publ., Dordrecht, p. 1-11.

Have, H. Ten (1995), *Approcci europei all'etica nella medicina clinica*. In: Viafora, C. (ed.): *Comitati etici. Una proposta bioetica per il mondo sanitario*. Fondazione Lanza-Gregoriana ed., Padova.

Have, H.A.M.J.ten and Lelie, A. (1988), *Medical ethics research between theory and practice. Theoretical Medicine and Bioethics* 19: 263-276.

Have, H.A.M.J. Ten, (1998a), *Health care and the human body. Medicine, Health Care and Philosophy* 1(2): 103-105.

Have, H.A.M.J. Ten (1998b), *Images of man in philosophy of medicine*. In: Evans, M. (ed.): 1998, *Critical* reflection *on medical ethics*. JAI Press, Stamford (Conn.), p.173-193.

Hope, T. (1999), Editorial: *Empirical medical ethics. Journal of Medical Ethics* 25: 219-220.

Hunter, K.M. (1988), *Making a case. Literature and Medicine* 7: 66-79.

Inglehart, R.(1990), *Culture shift in advanced industrial society*. Princeton University Press, Princeton, N.J.

Jensen, U.J. (1987), *Practice and progress. A theory for the modern health-care system*. Blackwell Scientific Publications, Oxford.

Jensen, U.J. (1989), *From good medical practice to best medical practice. International Journal of Health Planning and Management* 4: 167-80.

Jonsen, A.R., Siegler, M., and Winslade, W.J. (1986), *Clinical ethics: A practical approach to ethical decisions in clinical medicine* (2nd ed.). Macmillan Publ. Comp., New York.

Jonsen, A.R. and Toulmin, S. (1988), *The abuse of casuistry*. University of California Press, Berkeley, CA.

Jonsen, A.R.(1990), Practice versus theory. *Hastings Center Report* 20: 32-4.

Kass, L.R. (1990), *Practicing ethics: Where's the action? Hastings Center Report* 20: 5-12.

Kuczewski, M. G. (1997), *Fragmentation and consensus. Communitarian and casuist bioethics*. Georgetown University Press, Washington, D.C.

Ladd, J. (1983), *The internal morality of medicine: An essential dimension of the patient-physician* relationship. In: Shelp, E.E. (ed.): *The clinical encounter*. D. Reidel Publishing Company, Dordrecht, p. 209-231.

Ladd, J. (1989), *The good doctor and the medical care of children*. In: Kopelman, L.M. and Moskop, J.C. (eds.): *Children and health care: Moral and social issues*. Kluwer Academic Publishers, Dordrecht, p. 281-302.

Latour, B. (1987), *Science in action*. Open University Press, Milton Keynes.

Leder, D. (1988), *The hermeneutic role of the consultation-liaison psychiatrist. Journal of Medicine and* Philosophy 13: 367-378.

Leder, D. (1994), *Toward a hermeneutical bioethics*. In: DuBose, E.R., Hamel, R. and O'Connell, L.J. (eds.): 1994, *A matter of principles? Ferment in U.S. bioethics*. Trinity Press International, Valley Forge (Pennsylvania), p. 240-259.

Lelie, A. (1999), *Ethiek en nefrologie*. Dissertation, University of Nijmegen.

Murray, T.H. (1994), *Medical ethics, moral philosophy and moral tradition*. In: Fulford, K.W.M., Gillett, G.R. and Soskice, J.M. (eds.): *Medicine and moral reasoning*. Cambridge University Press, Cambridge, U.K., p. 91-105.

Newton, A.Z. (1995), *Narrative ethics*. Harvard University Press, Cambridge (Mass.)

Pellegrino, E.D. and Thomasma, D.C. (1993), *The virtues in medical practice*. Oxford University Press, New York/Oxford.

Pellegrino, E.D. (1995), *The limitations of empirical research in ethics. Journal of Clinical Ethics* 6: 161-162.

Poirier, S. and Brauner, D.J. (1988), *Ethics and the daily language of medical discourse. Hastings Center* Report 18: 5-9.

Svenaeus, F. (1999), *The hermeneutics of medicine and the phenomenology of health. Steps towards a* philosophy *of medical practice*. Linköping University, Linköping (Sweden).

Taylor, Ch.(1989), *Sources of the self. The making of modern identity*. Cambridge University Press, Cambridge.

Tongeren, P. Van (1988), *Ethiek en praktijk. Filosofie & Praktijk* 9: 113-27.

Tongeren, P. Van (1994), *The relation of narrativity and hermeneutics to an adequate practical ethic.* Ethical *Perspectives* 1(2): 57-70.

Tronto, J.C. (1993), *Moral boundaries. A political argument for an ethic of care.* Routledge, New York/ London.

Vandenbroucke, J.P. (1990), Medische ethiek en gezondheidsrecht; hinderpalen voor de verdere toename van kennis in de geneeskunde? *Nederlands Tijdschrift voor Geneeskunde* 134: 5-6.

Weisz, G. (ed.): 1990, *Social science perspectives on medical ethics.* Kluwer Academic Publishers, Dordrecht/Boston/London.

Welie, J.V.M. (1998), *In the face of suffering. The philosophical-anthropological foundations of clinical* ethics. Creighton University Press, Omaha (Nebraska).

Williams, B. (1988), *Consequentialism and integrity.* In: Scheffler, S. (ed.): *Consequentialism and its critics.* Oxford, p. 20-50.

Winkler, E.R. (1998), *Applied ethics, overview.* In: Chadwick, R. (ed.): *Encyclopedia of Applied Ethics.* Volume 1. Academic Press, San Diego, p. 191-196.

Wulff, H.R. (1988), *Contemporary trends in healthcare ethics.* In: Have, H.A.M.J.ten and Sass, H.-M. (eds.): Consensus *formation in healthcare ethics.* Kluwer Academic Publishers, Dordrecht, p. 63-72.

Zaner, R.M. (1988), *Ethics and the clinical encounter.* Prentice-Hall, Englewood Cliffs, N.J.

Zussman, R. (1992), *Intensive care. Medical ethics and the medical profession.* The University of Chicago Press, Chicago/London.

Zwart, H. (1995), *Weg met de ethiek? Filosofische beschouwingen over geneeskunde en ethiek.* Thesis Publishers, Amsterdam.

Zwart, H. (1998), *Medicine, symbolization and the "real" body - Lacan's understanding of medical science.* Medicine, *Health Care and Philosophy* 1(2): 107-117.

PART 2.1

THE GOALS OF MEDICINE IN RELATION TO THE SUBJECTIVISATION OF HEALTH AND RATIONALISATION OF HEALTH CARE INSTITUTIONS

ROBERTO MORDACCI

MEDICINE AS A PRACTICE AND THE ETHICS OF ILLNESS

A MORAL UNDERSTANDING OF MEDICINE

Current developments in the biomedical sciences are going to produce profound changes in the nature of medicine in the next future. One can recall the immense revolutionary power of molecular medicine in relation with the way diseases will be treated and prevented once the Genome Project is completed and once the knowledge thus obtained is more and more extensively applied to clinical conditions: not only diseases with a genetic basis will have a clear diagnosis, but other illnesses will benefit as well from treatment deriving from the use of recombinant DNA techniques. The present image of medicine as a mainly surgical and pharmacological enterprise will be profoundly modified by the increasing presence of a biotechnological drift, centred on cells and particularly on the works of DNA[1]. This change represents a dramatic shift in the ability to fight diseases, since in most cases we will be able to radically remove the cause of a disease (a gene) or to modify the genetic endowment so as to make an organism resistant to a specific disease. In the face of these changes, contemporary culture is called to rethink the nature and role of medicine, although such a reflection is anyway needed today in order to offer a better interpretation of the relation between medicine and other forms of human activity. The basic question seems to be: what should we expect from medicine as a part of the human effort to relieve the condition of suffering and to enhance the prospects of a flourishing life? In other words: will molecular medicine improve our ability to cope with illness and suffering, and in general with the limits of the human condition, or will it exacerbate the sense of alienation that so many people feel today in hospitals and in the course of some medical treatments?

The interpretation of medicine needed in such historical and cultural circumstances is not just an inquiry on its epistemological status, as if medicine could be defined only by the kind of knowledge on which it is based: neither a philosophy of medicine as a speculation on medicine, nor as a logic of medicine, nor as a philosophy of the medical science[2] would serve our present cultural needs in this field. In this sense, the growing evidence that both biomedical practice and research intrinsically raise moral as well as legal, political, social and economic problems makes it clearer than ever that to think of medicine as a science or simply as a scientific activity, although a very complex one, is not only reductive but simply wrong. On the other hand, many physicians still tend to interpret themselves as scientists, and the scientific foundations of their practice is being repeatedly stressed to gain public recognition and authority. Today we can better see that the modern

C. Viafora (Ed.), Clinical Bioethics. A Search for the Foundations, 101-113.

and contemporary interpretation of medicine as a part of the scientific endeavour is less adequate than the ancient and medieval interpretation of it as an "art", even if this term does not cover the actual status of medicine nowadays. It cannot be a way out of this impasse just to say that medicine is at the same time an art, a profession, a science and a commodity[3].

What we rather need today is a comprehensive understanding of the nature of medicine that makes it possible to see directly how the moral aspects of medical practice can be understood as inherent rather than juxtaposed to it; this means that the nature of medicine must be looked at from an explicitly *moral* point of view. Probably, the best way to do this is to try to understand medicine as a *practice:* that is, a kind of human activity whose fabric is a texture of meanings and goals expressed in the various actions that constitute it as a recognisable human endeavour. In this perspective, we should ask what kind of practice medicine is, i.e. which meanings, goals and values constitute the nature of this specific human activity. An interpretation moving along this line can open the space for an ethic of medicine that springs from inside it, and is not imposed on it from the outside, e.g. from an ideology, a philosophy or a religion; and yet, as we shall see, medicine is not an isolated practice, but it is a part of the pervasive search for the good that constitutes human agency, and it therefore receives its sense in a strict dialectic with all the other practices human life is made of; an "internal" interpretation of the moral sense of medicine does not entail a separated special morality disconnected with the rest of the moral life.

LINES OF A PHILOSOPHY OF MEDICINE

Philosophical interpretations of medical practice have already been developed by many influential scholars, whose reflections already clearly point in the direction above suggested: probably the most thorough of these attempts is the philosophical interpretation of medicine offered by Pellegrino and Thomasma in their *A Philosophical Basis for Medical Practice.*

The line of the argument developed by these authors highlights the essential features of medicine as a practice, although with some ambiguities concerning the very use of the term "practice"; my own contribution is an attempt to clarify this point with a more precise definition of practice and, therefore, a more precise definition of the goals and values defining this kind of practice. The characterization given by Pellegrino and Thomasma starts with some insights deriving from a phenomenology of the clinical encounter. For these authors "a specifically medical event is formed by clinical interaction" and

> "the moral nature [of medicine] stems from the fact that patient and physician mutually enter into a healing relationship"[4];

the modes of this relationship (i.e. what is important in it - morally important and relevant for its definition) are the characteristic traits of *responsibility, mutual trust, decision orientation* and the *curative intention* (what the authors call the "aetiology" of medicine). These traits represent, though rather syntethically, the result of a phenomenological consideration of the clinical encounter: they emerge as formal characteristics of the experience lived by the individual consciousness involved in the clinical relationship.

Pellegrino's and Thomasma's analysis catches some of the relevant dimensions of medicine as a practice. Yet, at closer scrutiny this phenomenological analysis does not seem to be taken to its ultimate level: the experience behind the clinical encounter - that is, the experience of illness, the need for help (for a specific and competent kind of help) - and the practical and verbal answers offered by the presence and the activity of the physician, the nurse and the healthcare system, generate a kind of relationship which is not exhausted by the formal and rather exterior modes of responsibility, trust, decision and intention. The most basic feature of that encounter is the *quality* of the relationship that takes place during its course. In that relationship, a whole universe of discourse is at stake, one in which illness should be given an understandable and credible sense which patient and caregiver can possibly share, at least as a background to their practical exchange. An adequate understanding of the implications of the phenomenological analysis of the clinical encounter leads to the discovery of a deep and intrinsic connection between the technical activities and the scientific knowledge of medicine, on one side, and a simpler and more radical form of life on the other, a form which is at the basis of the human relationship with every other reality - being it a person or any other entity in the world. That original form of relationship *is care,* and medicine represents a specific and highly sophisticated determination of that form: a form in which care is the competent and scientifically based practice of caring for the ill.

Although phenomenological approaches are not so common in the philosophy of medicine, phenomenology is certainly the most effective method, and the closest to personal reality, in order to try to understand what medicine is for the conscience of human beings (phenomenology claims to be, if anything, a method for the conscience to bring to clarity the original nature and content of its experiences). Merely conceptual analysis would not serve the aim of rendering the essence of medicine, since such an inquiry should first of all concern those contents of the experience of the healing relationship relevant for the individual consciousness and not a formal definition derived from an abstract concept. What really is at stake in the medical practice is the symbolic exchange effected in and through the actions performed by physicians and patients, and recognised by the cultural context as having a specific meaning.

Pellegrino and Thomasma argue that

> "medicine clearly is a domain of activity which is distinctive and distinguishable as science, art, and *praxis*. It comprises a set of legitimate philosophical issues and questions which derive from the unique nature of the clinical encounter"[5].

Yet, the characterization of medicine as

> "a distinct intermediate discipline, a tertium quid, between art and science but distinct from both of them [...] a habit of practical understanding refined and perfected by experience in dealing with patients",

or as a

> "derived discipline, a third class of human enterprises that combines theory and practice in a unique way"[6]

is unsatisfactory: it would be quite strange to recognize that we have no single concept to describe, from a general point of view, the kind of human activity that medicine is. Being a third class of human activity would make of medicine an odd discipline, as if, through centuries of history and dramatic mutations, we were not able to recognize medicine as an activity with distinctive characters; on the other hand, it is also true that many other contemporary forms of activity could be said to occupy a middle ground between science and "art" in the general sense of a practical enterprise (the truth is that we cannot use any more the term "art" as the medievals did), but it is unclear what such a positioning means from an epistemological point of view. More fundamentally, such a characterization does not clarify how theory and practice interact in the domain of medicine: what is the role of scientific knowledge in an art? And is there any possible unity of vision between the points of view of the physician as scientist, healer (artist) and caregiver (acting practitioner)? How could we ever avoid the possible conflicts between these different perspectives?

In fact, Pellegrino and Thomasma, after their analysis of the modes of the clinical encounter and of the forms of medicine, define the basic form of medicine as that of

> "a kind of craftsmanship of healing placed within an imbalanced relationship", a craftsmanship that "involves healing the body with the body"[7].

In this sense, one can say that the fundamental essence of medicine lies in its *practical* dimension, where theoretical knowledge serves a specific goal (that of healing) in the context of the clinical relationship. Yet, the two authors do not explicitly endorse a definition of medicine in terms of practice, although the use of the term "craftsmanship" points in this direction and medicine is often spoken of or referred to as a practice (even the title of their book bears the expression "medical practice" rather than simply "medicine").

MEDICINE AS A PRACTICE

One might suggest that medicine could fall under the domain of what Aristotle called *practical science*[8] but such a characterisation would be difficult to handle under the contemporary prevailing interpretation of what "science" means[9]; furthermore, this characterization bears some ambiguity about the status of the discipline, since the latter is defined essentially as a "science", that is, a theoretical activity, although with a practical aim: this may be true of medical knowledge (and therefore of the biomedical sciences nowadays) but it is false as a description of medicine identified with medical practice: the mere existence of treatises of medical knowledge would not suffice to determine the existence of medicine; such formalized knowledge may even be not strictly necessary for an effective medical practice.

On the other hand, there is no doubt that, while Aristotle defined medical knowledge as practical science, he repeatedly ranked medical practice among the arts *(technai),* especially in examples and analogies[10]. Yet, if we look at the definition of *techne* given in book VI of *Nicomachean Ethics,* we cannot avoid feeling a bit uneasy at applying such a definition to medicine (and not only to modern but to ancient medicine as well). Art is defined as

> "a state of capacity to make, involving a true course of reasoning" and "there is neither any art that is not such a state nor any such a state that is not an art" (the given example is architecture)[11];

moreover

> "the reasoned state of capacity to act is different from the reasoned state of capacity to make. Hence too they are not included one in the other; for neither is acting making nor is making acting"[12].

There are some passages in which medicine is presented as analogous in some respect with

> "matters concerned with conduct and questions of what is good for us"[13],

that is, with matters related to *praxis* rather than *techne*, but on the whole the caracterization of medicine as an art remains quite clearly prominent in the Aristotelian text.

Nonetheless, there are reasons to doubt that the Aristotelian characterization of medicine as an art can really cover what we call medical practice. This kind of thought especially raises when we think of the supposed *telos* of the "medical art", that is health: to be short, it seems too reductive to think of the health of a person exclusively in terms of a product, the result of a "making" *(poiesis).* First, health is a multifaceted notion, and although on the physical level of the concept it might seem appropriate to say that health is the result of, say, an appropriate medical treatment, we nonetheless know that in many instances a state of health means something more complex than that, i.e. a condition sometimes compatible with an impairment. Second, health is normally not the result of any medical treatment, but a normal state of the individual which is threatened by illness and which medicine tries to restore; so it is not, strictly speaking, a product of the medical art (as, on the contrary, a house is the product of the making of architecture), but a result of the dynamic interaction between the natural processes and the healing operations of the physician. Third, longing for health is a far deeper desire than just the aspiration to physical well-being, as it appears from phenomenological description of the emotions and expectations of patients facing illness (no matter how serious)[14]. Fourth, health is a good that receives at least part of its meaning and content from the relationship between the healer and the healed, the caregiver and the cared for, a good that is partially constituted by the symbolic exchange of meanings which takes place in the clinical encounter[15]. Fifth, health is a relevant feature of the overall good of persons and therefore contributes, though in a complex way, to happiness, so that to be concerned with the health of persons means at the same time to be concerned with what is good for them; at least, it is always important when deciding about a medical treatment to see it in the light of the more general good of the patient[16]. Finally, what often seems important in medical practice is not so much the goal of health but the act of *healing* or at least of *caring,* which is in some respect independent from the achievement of that goal: what a patient expects from a physician and a medical team is first of all to be cared for, with the hoped result to be healed;

but the fact that a medical team may not be able to restore the patient's health does not deprive their actions of their quality of *medical* actions; essentially, all that the medical staff does is an expression of the fundamental commitment of medicine, that of caring, even if the goal of health cannot be achieved.

These considerations strongly suggest that medicine has much in common with the realm of *praxis;* even, we might say that, since its aim is a feature of the good of the person and since that good is exchanged primarily in the context of an interpersonal relationship, the essential carachterization of medical practice could be sketched in terms recalling the Aristotelian notion of *praxis* ("matters concerned with conduct and questions of what is good for us") rather than that of *techne.* It is therefore easy to see that this considerations do lead us to refer to a notion which has received much attention in the recent debate in moral philosophy, i.e. the notion of *practice.*

An alternative way to characterise medicine, then, might be to consider it *a practice using a vast domain of knowledge for the purpose of healing.* A general definition of practice, although quite a complex and long one, is that given by Alasdair MacIntyre, designed in a loosely Aristotelian context. According to this definition, a practice is

> Any coherent and complex form of socially established cooperative human activity through which goods internal to that form of activity are realized in the course of trying to achieve those standards of excellence which are appropriate to, and partially definitive of, that form of activity, with the result that human powers to achieve excellence, and human conceptions of the ends and goods involved, are systematically extended[17].

To reduce complexity to a schematic list of properties, the basic features pointed out by this definition can be summarized as follows:

1. Coherence and complexity: a human activity constitutes a practice if it has some degree of complexity and internal coherence, i.e. if it is not just the result of spontaneous and incidental activity, pursued without any meaningful intention or made up of a mere sequence of acts belonging to different and irrelated practices. According to MacIntyre's examples, bricklaying is not a social practice, architecture is.

2. Cooperation: a social practice implies the cooperation of various agents for a commonly shared goal. A completely solitary activity, carried out by an individual with no recognition or mandate by others, does not constitute a practice.

3. Internal goods: the performance of a particular practice is determined by the attempt to realize a determinate figure or form of the good, which cannot be realized otherwise. To use MacIntyre's example, playing chess can lead to prestige and money, if one is a champion, but these are not goods intrinsic to playing chess; the internal values of playing chess can be achieved only in the very act of playing at the best level of excellence possible for the subject; the masters of a practice are those persons able to realize its internal goods at the highest levels (or even better, see below). The notion of internal goods is crucial for the definition of practice, as it permits to distinguish a particular practice from any other and allows to draw the important distinction between internal and external goods; the latter can be included

in the practice but they do not define its identity since they belong more appropriately to other forms of activity.
4. *Standards of excellence:* any subject performing a practice can achieve its goods in a more or less adequate manner, but there are standards of excellence that define the minimal socially accepted levels of correspondence between the performed activity and the achievement of its internal goods; performing at the basic standard of excellence is required to enter the relevant circle of those performing a certain socially recognized practice, but it is implicitly assumed that any one entering the field of that practice will try to improve his ability to achieve its internal goods.
5. *Extension of human powers:* the goods internal to a practice can always be realized better; thus, there is an open space for those who have acquired a special dominance over the virtues necessary to realize those goods, so that they may achieve them in an unprecedented degree in the history of that practice. Their enterprise can be counted as establishing a new frontier in the capabilities of man concerning that practice.

It is perfectly clear, as also Jeffrey Stout has noted[18], that medicine is a social practice according to MacIntyre's definition: medical care is a complex and coherent activity, which implies cooperation between various people pursuing a common general good through different individual tasks; this common general good can be said to be internal to medical practice in the sense that it cannot be achieved otherwise, so that any way of achieving that good (which we can assume is "healing") constitutes an example of medicine; there are standards of excellence below which nobody is publicly allowed to practice medicine; and the masters of the practice are those who extend its limits in realizing its internal goods.

Although MacIntyre's definition uses the more general notion of good rather than that of goal, we might assume, in this context, that the senses of the two terms largely overlap: a goal (or a *telos* in the Aristotelian sense) is the good pursued in a particular kind of action; there are goods that are not the result of an action - so that they do not constitute a goal for human action - but human agency is characterized by the tendency to achieve a goal which is understood as a good by the agent. In the following, we will rather take up the notion of goal, because it is more determinate and because the discussion concerning the nature of medicine usually occurs with reference to its presumed goals.

In a more analytical style, Lennart Nordenfelt has defined medicine along similar lines, characterising it as

> "the practice performed or supervised by physicians/ psychiatrists in their professional activity of enhancing health by preventing or treating diseases, injuries or defects"[19].

It is visible that the centre of this definition is still the patient-physician relationship, although in an indirect sense, since the focus here is on the activity of those who profess to heal (physicians/psychiatrists and those supervised by them); what matters is what those people do in the attempt of enhancing health - a goal that includes prevention and treatment. Therefore, this seems a good definition, although problems persist with reference to the notion of health as an intrinsic goal of medical practice.

INTERNAL AND EXTERNAL GOALS

Of course, to describe what kind of practice medicine is, it is necessary to identify at least one distinctive internal good, by virtue of which all the activities performed under the general label of medical practice are recognizable in a coherent manner. External goods are those goods which can be achieved or realized in the course of a particular practice, but are not peculiar to it: for example, medical practice usually brings money, power and prestige to physicians, but these are external goods, not goods that are peculiar and distinctive of medical practice.

Before trying to specify the good internal to medical practice, a question must be answered concerning the nature of that good. Although the distinction between internal and external goods proposed by MacIntyre is, as Stout recognizes, "tolerably clear"[20], some may argue that there is no such thing as an internal good to be pursued as a goal distinctive of a particular practice: any social practice receives its goals (and its goods) from the cultural context in which it is performed, and there is no "essential" goal or *telos* that objectively defines a practice all around the world and in any human society; what we, in the industrialized West, call medicine is something quite different from what wizards and magic healers do in other cultures or even from what Westerners used to do in Ancient and Medieval times. Therefore, so the argument goes, the distinction does not lead to the identification of any hard core values of any practice; rather, an "internal" value has to be recognized, from a distance, as a value imposed on the practice by the surrounding culture; the variability of the goals that identify medicine in the different times and places implies that those goals are not *intrin*sic, but simply a social construction; one can say that

> "medicine is thus best thought of as an evolving fund of knowledge and a changing range of clinical practices that have no fixed essence"[21].

In particular, the key concepts of health and illness appear to be too strictly dependent on the cultural context to be used to define a practice throughout the world.

The question concerning the nature of the goals defining a practice seems to be a crucial one, in particular for medical practice, since the ethics of medicine depend on which values are considered the proper ones for it as a practice, and those values depend on the goals defining medical practice. To start with, I think that the alternative external-internal should not be overstressed: any form of practice has internal values, but they are embodiments, or practical examples, of values that are *always at the same time to a certain extent belonging to the general existence of human beings;* a practice cannot be totally severed from the wider texture of human existence, for it is in our practices that we are trying to give reality to our hopes and our beliefs concerning the whole of our lives. Thus, there can be incompatibility between the particular *telos* of a practice and the general human search for the good only if we have already lost the original connection between them, that is, if we have separated the particular practice from the search for the good life as a whole. This comprehensive search constitutes the texture of the fabric in which the particular practices can have an intelligible sense; and this connection between the overall search for the human good and the goods internal to a practice constitute precisely its *moral form,* that is, its meaning as a part of the moral life of an individual or a society. There can be an excellence in purely technical terms, a skill of performance which realizes at a very high level the goods internal to a determinate practice, but it may well be the case that the person with such an ability loses sight of the place held by the values inherent to that practice with respect to other values that are more general or closer to the most relevant dimensions of human life.

Any human practice aims at realizing goods that are modes of the overall notion of the good life, and there is no practice which can be said to be all-encompassing as a way to realize the good life. It is for this reason that the traditional cardinal virtues can be traced in most of human activities, since, as MacIntyre himself recognizes at least for justice, courage and honesty, they are qualities that need to be cultivated for any kind of activity.

Furthermore, we have good reasons to say that the goals of medicine are certainly subject to deep variations among the cultures, but that they nonetheless make medical practice recognizable everywhere: it is not arbitrary to define a certain practice as belonging to (a form of) medicine, although the range and scope of what it entails is heavily dependent on the particularities of a specific culture. The basic feature of any of these activities is the attempt to *care for an ill person,* by any coherent definition of illness and by any means one can make use of.

Therefore, the sense of a practice can be said to be socially construed but not on empty spaces: medicine is a practical response to the challenge of illness in terms of a scientifically sound and technically developed form of care; this practical response takes a number of different forms depending on the way a particular culture in general considers the event of illness and the ways to face it from a moral and from a practical point of view. These meanings need not be reciprocally incompatible but can vary significantly while remaining under the general intention of caring in a competent manner; the concrete form of a competent act of caring is obviously dependent on the scientific development of a determinate culture or historical period and therefore variation is nothing strange in this general characterization of medical practice; furthermore and finally, medicine aims at restoring or enhancing health, but at the same time it recognizes that this may not always possible and that it can assume the less ambitious aim of palliating or assisting.

Modern medicine is more technical than practical; but the essence of the medical enterprise is still a practical one; we would not recognize as medicine a practice completely detached from the human effort to give meaning to illness and suffering, and yet the question of the determination of an understandable sense of illness is a larger goal than the one that defines medicine; medicine is limited to a *practical* response, that of caring in a competent manner, i.e. using the best resources of scientific knowledge. The elaboration of a sense for illness, suffering and the desire of health is not a goal of medicine, but *physicians cannot practice medicine without appealing to a shared sense of illness and health,* because the dimensions of illness and health cannot be reduced to the merely biological; their analogical complexity calls for a deeper awareness of this multilevel meaning.

> "The reality that counts is cultural reality, and the system used by the healer or doctor need be accurate only in terms of the culture in which it is being used, for it serves to explain illness"[22].

Thus, we can define medicine as the practice of trying to heal or rather the practice of competent care for the ill, because, whatever health and illness may mean, they will nonetheless identify a practical goal for all those involved in the situation of meeting human suffering and having at least the presumption to be able to heal. In this sense, it can be said that medicine is a form of caring that aims at curing and that healing is its ultimate normative ideal but not a necessary result.

Care, cure, healing and health enhancement have their sense in the relationship of proximity with the suffering other; they represent the whole spectrum *(at least care for the ill, at best heal or, even, enhance health)* of the *telos* of a practice whose technical aspect has become prevailing in the contemporary age, obscuring its roots in the practice which originally gave the meaning to the technique. It is in the practice, that is in the relationship, not in the technique *per se* that the exchange of meanings takes place, that is, care occurs: so to say, not the skilful gesture of the surgeon *per se is* medicine (it could be an act of slaughter as well), but that act together with its sense as a moment of the caring relationship; only within the horizon of the caring relationship that gesture constitutes a medical act.

Every culture defines the contents of medical practice in ways partially different from that of other cultural worlds, starting from different interpretations of the experiences of illness and health; these experiences, in particular that of illness, have traits that are originally common to all humans (they constitute the original core meaning inscribed in the experience of illness)[23] , even if they can be construed in significantly diverging ways, to the point that they give rise to quite different forms of practice as a competent response to illness. This does not mean that we cannot recognize medicine in the diversity of its historical forms, because we are authorized to call medicine any concrete form of caring relationship aimed at healing which uses the means and ways considered to be most adequate in a specific cultural context. Such a characterisation implies an analogical concept of illness and health, which allows to recognize the persistence of identity through variation in the different concrete embodiments of the historical notions of illness, health and medicine.

RETHINKING THE ETHICS OF ILLNESS AND HEALTH

One of the necessary steps in order to rehabilitate the identity of medicine as a practice is the rethinking of the ethics of illness: illness is in fact an event which calls the individual's conscience to a special challenge, that of defining his or her personal way to resist illness. The ways of this resistance are elaborated by the subject on the basis of his or her cultural and moral resources and they in turn contribute to the determination of the style of a certain culture with reference to the ultimate questions of moral life.

Medical practice is inscribed in this movement, and it can significantly determine its direction; if, as our culture is presently doing, we deliver to the technological side of medicine all the mission of facing not only the material challenge of illness but also the challenge of its meaning for us as human beings, we implicitly endorse an understanding of illness as devoid of any sense apart from the pure idea of negative: in fact, as a purely technological enterprise, medicine cannot give any other meaning to fatal illness than that of "failure". Thus, if medicine is seen as a technical answer to a technical problem, its only rule is efficiency; and yet, this path takes us away from any possibility of giving any sense to all the situations in which we cannot heal nor cure, but only care; the result is that this situation leaves the patient (who is declared incurable or terminal) completely alone. The main reason for the dissatisfaction of many people confronting medical practice lies in the absence of a perceived framework of meaning sustaining and animating the technical operations doctors and nurses perform at the bedside; the language and the gestures of contemporary medicine are not viewed as communicating an understandable sense for the condition of illness, in particular when the latter is chronic, degenerating or incurable. Medicine needs to be connected through a network of meanings to a wider perspective in the moral life, where competent care is a part of the larger effort to face the limits and the promises of the human condition. The moral resources designed to give a sense to illness, suffering and death, together with the practical answer of caring and the activity of the healthcare professions, constitute what we might call the *ethics of illness:* not a set of norms but a framework of sense which is made present in the act of choosing to realize certain goods in the face of the challenge posed by illness.

Another side of the problem is the question of enhancing, and this is an aspect we can only mention here. Today we are facing the challenge of finding a reasonably clear sense of what it means practically to promote or enhance health, since we are ever more capable of doing it. We might call this side of the problem the *ethics of health,* in the sense of a moral reflection concerning the potentialities of contemporary medicine. The bioethical discussion on this issue is still quite underdeveloped and it is strictly connected with the debate on the prospects of human gene therapy. A very general criterion in this area would be the notion of "human nature", if only this concept were not so despised by many contemporary philosophical perspectives[24]. Anyway, such a concept seems one we cannot do without, as even the critics sometimes admit[25], and we should try at least to suggest some essential traits of human beings in order to have a normative guide.

The widespread interpretation of the notion of human nature is that of a fixed reality whose boundaries constitute normative barriers against the attempt to modify the present existential conditions of human life. Needless to say, this is a rather materialistic (and indeed biologistic) understanding of human nature, even if, as it seems, it has been assumed and defended by many religious perspectives in the first place. A more dynamic understanding of human nature is nonetheless implied by those who recognize at least two essential traits of human beings: *rationality* and *freedom;* usually these two traits serve also as normative guides, in the sense that, for example, the detriment of the rational abilities or of the free will of an individual by means of any medical or technological intervention is considered unethical; even an enhancing project, if it implies the risk of diminishing the functional ability of man as a rational and free agent would be considered unacceptable in this perspective. I would add to these traits that of *having (or rather being) a recognizably human body:* man is an embodied rationality and a culturally and historically situated freedom, or, in other words, is *a bodily self.* For a human individual it is essential to have a body[261], in the sense that the personal characteristics of the body may change, even dramatically, but if I distort the human body I am distorting the human self and therefore I am not enhancing health.

Thus, enhancing supposes that we consider the interrelation between faculties and powers: the harmony of the human being is an equilibrium which can be redesigned continuously (as we have been doing throughout history and not only in medicine). The problem is that very often we do not know enough to predict the outcome of an intervention in one or the other of the dimensions of human life, since every dimension is so strictly connected with any other that a small change 'm one part may bring to the disruption of a delicate balance in another. This warning calls not for inaction, but rather for prudence, that is, for the exercise of discernment in front of any single choice posed by the potentialities of introducing changes in the features of our present condition.

Roberto Mordacci, Associate Professor of Philosophical Anthropology and Bioethics, University "Vita- Salute", San Raffaele, Milano, Italy.

NOTES

[1] For an overview of these developments see L. Walters J. G. Palmer, *The Ethics of Human Gene Therapy,* Oxford University Press, New york, 1997.

[2] For these form of the philosophy of medicine see H.T. Engelhardt, K. W. Wildes, "Medicine, Philosophy of", in W. T. Reich (ed.), *Encyclopedia of Bioethics,* Macmillan, New York, 1995, pp. 1680-1684.

[3] Cf. E. D. Pellegrino, D. C. Thomasma, *A Philosophical Basis of Medical Practice,* Oxford University Press, New York, 1981, pp.196-199.

[4] Ibid. , p. 63 and 64.

[5] Ibid. , p. 25.

[6] Ibid. p. 59 and 61.

[7] Ibid. , p.72 and 73.

[8] Cf. E. G. Aristotele, *Metaphycs,* A 981a-982a. The basic idea is that a prctical sciences is a formalized (rigorous and objective) cognitive activity with a practical aim.

[9] The modern and the contemporar idea of sciences entails not only the idea of true knowledge of the causes, but it is intrinsically related to a specific method (the observational-experimental one) designed to obtain rigorous and objective information. The point is that such a method is not always applicable to a practical knowledge and even less to practical activity.

[10] Cf. e. g. *Nicomachean Ethics*, I, 1094a; I, 1097a; II, 1104a; V, 1138b; VI, 1138b; VI, 1141a; VI, 1143b; VI, 1145b.

[11] *Nicomachean Ethics*,VI, 1140 a, (trans. D. Ross).

[12] *Ibidem*.

[13] *Nicomachean Ethics* II, 1104 a (trans D. Ross): "matters concerned with conduct and questions of what is good for us have no fixity, any more than matters of health. The general account beings of this nature, the account of particular cases is yet more lacking in exactness; for they do not fall under any art or precept, but the agents themselves must in each case consider what is appropriate to the occasion, as happens also in the art of medicine or of navigation".

[14] I tried to suggest some considerations in this direction in "The desire for health and the promises of medicine", *Medicine, Health Care and Philosophy*1 (1998): 21-30.

[15] In a roughly similar way, Pellegrino and Thomasma suggest that the health is a "relational good", cfr. *For the Patient's Good*, Oxford university Press, New York, 1988.

[16] This is also suggested by the characterization of the patient's good offered by Pellegrino and Thomasma in their *For the Patient's Good*: the comprehensive notion incluudes the "biomedical good", the "bst interest" of the patient, the "personal good" and the "ultimate good".

[17] A. MacIntyre, *After Virtue*, 2nd ed. , Notre Dame University Press, Notre Dame,1984, p. 187.

[18] J. Stout, *Ethics After Babel. The Language of Morals and their Discontents,* Beacon Press, Boston, 1988, pp. 266ff.

[19] L. Nordenfelt, "On medicine and other species of health enhancement", in L. Nordenfelt, P. Tengland (eds.), *the Goals and Limits of Medicin,* Almqvist & Wiksell International, Stockholm, 1996, pp. 33-49, p.40.

[20] Ibid. , p. 268.

[21] *The Goals of Medicine,* Special Supplement, *Hastings Center Report* 26 (Nov.- Dec. 1996), p. S7.

[22] Cf. A. MacIntyre, *After Virtue,* p.191; cf. J. Stout, *Ethics After Babel*,p. 269.

[23] Cf. R. Mordacci, " Health as an aanlogical concept", *Journal of Medicine and Philosophy* 20 (1995): 475-497.

[24] Cf. e. g. K Bayertz, *Gen Ethics. Technological Intervention in Human Revolution as a Philosophical Problem*, Cambridge University Press, Cambridge (MA), 1994.

[25] Cf. J Harris, *Wonderwoman and Superman: The Ethics of Human Biotechnology*, Oxford University Press, Oxford, 1992.

[26] Maybe not this body, but a body recognizable as human.

PAOLO ZATTI

THE RIGHT TO CHOOSE ONE'S HEALTH

PREMISE

There are juridical concepts which seem to be destined to lead the interpreter and the judge to the experimentation of the bounds of their own tasks and of the instruments entrusted to them: the limits of right and jurisdiction. The idea of health is one of these concepts.

The gift of health and the right to health function as a basis and represent a crucial point for the different aspects of the person's protection: the right to health care as a public service; the right to a healthy environment, in all its possible definitions, the right of compensation for damages, especially, those damages linked to medical responsibilities; the problem of consent for medical treatment.

The two last points, which particularly involve the expert in private law, are connected to each other. As to the cases of medical responsibility due to missed consent or missed information, what brings them together is fault.

The idea of health which involves and links both themes is acquiring contents of special topical interest within jurisprudence and the doctrine. At the same time, within the connection of responsibility-consent, they make the problem of juridical treatment of the -patient relationship closer to the above-mentioned boundary zone.

The key of this evolution - which makes the meaning of "health" richer and finer - is a different ideal and complex semantic connection, the one between health, the freedom of disposing of one's own body and personal identity.

HEALTH AND IDENTITY (MARGINAL TO SAN RAFFAELE'S CASE)

Health: a "fundamental right"

In our juridical horizon, the link between personality, freedom and health finds its own root in the article 32 of the Italian Constitution, which qualifies right to health as "fundamental". The adjective, inserted somewhat doubtfully into the final text and superseding a previous choice of preparative works in which the guarantee was formulated in a looser way, has the value we know: the qualification of right as fundamental inserts health among the rights which must be ascribed to individual, since those rights are inscribed in the concept of individual and of this concept itself they form the main features which cannot be disregarded. The constitutional protection of health is linked, therefore, to the general protection of personality (art.2), to the fundamental rights of equality and dignity (art. 3), and of individual (art. 13), on the one hand, and to the norms which guarantee the social

C. Viafora (Ed.), Clinical Bioethics. A Search for the Foundations, 115-129.

manifestations of the subjects, especially, their family roles, their insertion in educational and instructive processes, insertion in the work-field and in social life[1].

The connection between health and freedom becomes of particular significance in the grounds for the judgment which constitutes the starting point of these pages; the Milanese judge refers to the decision no. 471/90 of the Constitutional court (In Giur Cost., 1991, 626 f. with a note by MUSUMECI and in Foro it.-, 1991, I. 14f..; with a note by ROMBOLI) which inscribes the freedom of disposing of one's own body in the individual freedom guaranteed by art. 13; this means to state not only a strong principle of self-determination about governing one's own health but also establishing an interaction at a level of content between health, freedom and personality, as will be shown later[2].

The definition of "health"

These "systematic" links, so familiar to the Jurists that they are taken for granted, represent, however, an interpretative instrument producing "enormous" results when a dynamic element i.e. the idea of health is inserted.

The Constitution does not define health. Undoubtedly, a fundamental right such as health asks for a content proper to the idea of person conceived as unity of all those aspects, which the mind, with its reasoning by distinctions and contraries, observes and classifies as "components" of the person; particularly, the body and psyche.

This unity characterizes the contemporary definition of health stated in art. 1 of Constitution by OMS in 1948 and considers expressly the three aspects of well-being: health is the "complete state of physical, mental and social well being", to which the adjective "spiritual" was added at the end of the 80ies[3]; that is why health does not "merely consists in the absence of an illness or infirmity".

The definition by OMS is often considered with conceit or annoyance within the medical area. One observes that the definition is so wide and without limits that it is liable to any assertion placing the idea of health within an area of mere subjectivity. Therefore, one could add controversially that it is beyond the reign of the observable while from the point of view of useful guidelines to health policies, it is clearly considered as an utopia[4].

A scientific or epistemological criticism on the definition of health is not our task. Because of its obvious importance, one must consider, from a juridical point of view, that the definition by OMS is stated on an international normative source[5]. The definition of health as "a state of physical and psychic well-being in the Italian code of medical ethics (art.3, 2nd paragraph), though of lesser importance, is open to the unity of person, too.

The problem, with or without the basis of the normative definitions, resides in linking together two evolving aspects of the idea of health: a) the one by which it enlarges itself to the psychic, or mental, aspects of well-being and b) the one for which the well-being is centred on subjectivity, on the experience each individual has of its own physical, mental state and - in case of other states - in connection with the mental-spiritual one.

It is this concatenation which is highlighted in jurisprudence and finds its own mature and lively expression in the Milan Court's judgment.

Thei idea of health: utopia in jurisprudence?

The idea of health stated in jurisprudence is according to the "utopia" by WHO as regards two aspects: the reference to the unity of the person, on the one hand; the denial of correspondence between health and absence of illness, on the other. However, it adds to the definition of health a considerable "added value", which starts a small Copernican revolution : health centred on self-perception.
In 1988 (judgment no. 208, in Quadrimestre, 1988 433 f. with note by NATUCCI) the Court of Cassation establishes the principle according to which health is not a "mere organofunctional eurhytmics (i.e., absence of illness) but a

> "state of complete well-being involving the inner aspects of life as perceived and lived by the individual him/herself" (compare Cass. no.411/1990 in Orient. Giur. lav., 1991, 2387).

The most immediate consequence of such a re-definition is the insertion of negative emotional facts, not only among the causes detrimental to health (physically understood) but also among the events which alter the state of psychic well-being and which, therefore, represent in themselves *detriment* to health[6].

The causal relationship between psychological state and the fall of body's defences was stressed, in jurisprudence in Massimo's case (Assise App. Firenze, 26.6.1991, in *foro it.*,91, II, 236). In that case, the lack of information was in its turn considered as aggravation to the noxious psychological reaction. The fact of considering a negative emotion as a decrease of health in itself, -since being a decrease of the psychic well-being - opens more interesting perspectives, though very delicate both in the juridical and ethical field. Is pain in itself a loss of health? Is humiliation in itself an injury to health? Is fear in itself a decrease of health? A wide, maybe unrestrainable, fusion is emerging between health and other "goods" which are traditionally the content of rights *different* from that of health: the right to dignity, the right to decency - the offence of which, injures or cancels the mental and social well-being of the injured subject.

Anyway, the meaning and consequences of the above-mentioned definition have a wider significance.

Health from standard to lived experience

The concept of health, traditionally relevant in the juridical - and non-juridical field, could be defined as the result of comparison between the psychophysical state of a person and a pre-determined psychophysical *standard:* the one which could be referred to as the image of the "healthy man". In its turn, the standard of healthy man allowed to give importance to each grade of gained or lost health within a wider comparison with the goods and the interests of individuals and collectivity, by converting the fundamental right into the measures proper to the social rights, on the one hand, offering a method of comparative evaluation useful to the solution of conflicts with other interests and "goods", on the other The body-mind unity, together with the idea of complete well-being, could still be limited to spread the spectrum of comparison, by bringing together within the standard all the attemptable aspects of "eurhytmics" - the latter, an expression of the Court of Cassation. The logic pattern is to this extent applied to a larger factual basis, but it would remain unvaried.

However, the stress given to the psychic well-being would naturally act as a carrier leading the concept of health beyond the sphere of the objective pattern of comparison.

The "mental" component of health opens the concept to "the inner aspects of life perceived and experienced by individual": or rather, to the events which do not directly belong to the sphere of the observable, but that could be introduced into it by means of communication, interpretation and empathy; such events are represented neither in the language of the medical science nor in the language of other sciences but in the language of relationship[7].

The consequences on the physician-patient relationship are clear.

In the sphere of a concept of health understood as a comparison between individual reality and the standard of eurhytimics, health is definable on the basis of criteria belonging to the medical science. The comparison is entrusted to the observable, the verification of which is under the physician's competence. The patient may report his symptoms; the act of selecting them as relevant or irrelevant is under the physician's competence; a space of self-determination for the patient opens, instead, during the phase of therapy, to give his/her authorization to the disposal of his/her body. At this stage too, the definition of the therapeutical purpose is under the physician's competence.

When the concept of health opens itself to the inner aspects of life as the individual perceives and experiences them, the foundation of the relationship is overturned. Health becomes a concept which, first of all, expresses a perception of self as a complete subject. This does not mean that the standard is insignificant, and that the physician's competence is not involved, but the evaluation of the patient's condition in terms of health is necessarily dialectical, and the conclusion is possible only as a synthesis, as far as the medical thesis and the antithesis of self-perception are liable to be overcome and brought together by an agreed solution: beyond that, there is the patient's freedom which could be expressed with the denial to the therapeutical proposal, and opened to the problem of the prosecution of the relationship on a conflicting basis with the consciousness of the healer[8].

The physician-patient relationship is characterized, in such a manner, by a more complex "law of listening" which involves primary regard to the patient's self-perception as an ill individual; and which obviously interferes with therapeutical choices under the perspective of the purposes considered desirable[9].

The consequences, however, go beyond the health-relationship. From a juridical point of view, identifying certain acts as injurious to health and ascertaining the existence and the entity of injury to health was based on comparison with the standard in terms of decrease: once again, on a somewhat relatively "objective" evaluation under the medical competence.

Health as a lived experience implies a shift towards criteria which stress the injury to personality, and, consequently, include comparison with the standard but do not finish with that.
In short, I overlook further consequences such as the different importance which could be given to the problem of aesthetic intervention and change of sex and artificial procreation: how can their therapeutical character be denied?

From the idea of self to the choice of health.

The whole issue which has been reported up to now, is welcomed, specified and stressed in the recent jurisprudence concerned.

The Court of Milan, in deciding about S. Raffaele's case - a case of diagnostic intervention without the direct consent of the patient, but of one of the family member - refers to a change of perspective consisting in the bestowal of significance on the psychological dimension of health by means of fertile grounds. The various consequences are stressed.

The first consequence, particularly relevant, is the distinction between the purpose of treating a pathology and the purpose of pursuing a better health for the patient. The former cannot coincide with the latter, and it is not sufficient in itself to justify the medical action.

This specification is only a fall-out of the predominant assumption concerning the entire motivation: the one according to which health is modelled on the personal and unquestionable life's expectations and involves the idea an individual has of him/herself. The thing, if I am not mistaking, means to direct health towards the right of any person to the free research and individuation of one's own identity. This principle has very strong consequences on the problem of consent. The relationship with one's own self is obviously very private; it cannot be substituted with surrogates. On this basis, it is unavoidable to state that the patient has a total right to information but also a duty to be open to information. Identity is not something which one can be freed from. It is surely something that can involve a confidential relationship with an appointed "protector". But also in this case, a substantial informative basis or a clear and widened provision for the most serious hypotheses must be taken into account. The problem of information becomes, then, in its turn, much more delicate, because it implies the comparison with the patient's person in view of his/her "inner aspects of life".

Everything could be debatable, but an important aspect can be certainly stated in the governing of the relationship: the inner aspects of life and the sense of one's own identity are not matter for informed consent forms.

The point to be stressed from a juridical point of view is still the entanglement of freedom, identity and health. The gift of health is based on the freedom of disposing of one's self or rather of one's own person in the physical, mental and social aspects which form the well-being, reachable under given circumstances. Health is not distinguishable for some aspects from the person, at least, as regards individual capable of deciding, who is the only one to be able to attribute to the right of health a concrete content on the occasion of therapeutical choices: to better show, therefore, among the different options, which one corresponds with one's own best health state.

It seems thus unavoidable that the evolution - and both Massimo's and S. Raffaele's case prove this -is directed towards the clear strengthening of the right of not being treated. If the perception of one's own identity and the unquestionable expectations of life lead a person to accept a pathology and its results or to desire for being left to die, then this idea of health does not contrast such options; it confirms them. This means that health and desire for life are not always congruent; they would be such according to a wide and suggested standard but not according to the intangible nucleus of self-perception which becomes the core of the ideas of freedom and health - under this perspective, one could seek "healing" which means acceptance of illness, or the last great "healing" of one person - or rather, the acceptance of one's own death[10].

THE PROBLEM OF MEASURE AND THE ROLE OF CORRECTNESS

At the beginning, the problem of limits of right were mentioned. In fact, by considering the judges' normative positions, anyone ' could wonder if they are liable to effective application for all aspects of the treatment- relationship. The question mirrors a greater problem : to convert the ethical problems into juridical norms and principles. In what extent is law able to gather the criteria of moral conduct, by covering their structures, and to provide the latter with the “material" power of law? To stick to our issue: this means to ask the physician to be open to the listening to the patient and to observe together with the patient his/her health condition. All this by starting from the patient's self-perception and by taking into account the meaning that the patient gives to his/her own identity as regards the relationships between health and illness, death and life - this is possible in moral terms, in so far a person is called to confront him/herself to the idea of well-being before the mirror of consciousness.

Also in these terms, a conscious physician could declare his/her consternation and his/her lack of preparation: how does one learn to listen to and to interpret the signs which allow the knowledge of other people's interiorities? Which level of knowledge of one's self and of clearness of one's self is needed to support decisions in which another person's self-perception is involved? Which training allows to catch and govern empathically that play of reflections which establishes itself between the patient's expectations and the physician's answer? Furthermore, which objective conditions - i.e. concerning time, place, privacy and closeness - can make these modalities of conduct feasible, also from an ethical point of view?

The shift from an ethical speech to a juridical one, add to these difficulties, the ones concerning the structure, the purposes and the manner of performing the juridical norm.

How can the "performance" requested to the physician be individuated compared to the uncountable variables of the patient's condition? It is necessary for general provisions: by which key and by which juridical semantics?

Would a norm have a meaning when it prescribes to the physician respectful attention, patience and even sweetness? One should abstain from translating juridical norms into the image of ideal physician . One should abstain from imposing to the family members behaviour in an affective relationship, as well; it does not concern the juridical matter and jurisdiction.

Law has its own semantics of duty and also of agreement with morality. It involves the model of the loyal and honest person, of the "good" father of a family. It prescribes assistance, presence, diligence and information. Its limit is maybe the idea of correctness.

The nucleus of the duty of correctness is loyalty: the concern of duty is the loyal and honest performance of the due treatment, and more widely, the honest and loyal behaviour accompanying treatment. Loyalty demands for transparence of conduct, collaboration between the parties, "protection" of the other party's interests and, obviously, information.

In terms of correctness, the "*duty of support*" by the physician finds its own juridical foundation, as an aspect of mutual collaboration - I mean that behaviour by which the physician participates in the search for truth of the ill individual, or rather the search for a way of accepting the truth, and in the development of the relationship with truth, without levelling one's own role -also independently of the point of view of the "correct" content of treatment -to the giving of mere and standardized information according to the time and the ways which disregard individual development.

This is what makes sense, at a juridical level, in prescribing,: the juridical norm needs measure, it cannot prescribe without measure. The parameters of behaviour and evaluation deriving from the principle of correctness are characterized by measure. The correctness is the sense of measure, of what one has to do and what one has to demand [11].

Measure and "purity" of the fundamental right

I would like to point out that the necessity for giving "measure" to the consequences of the concept of health defined by jurisprudence, and, especially, in the judge about S. Raffaele's case, does not mean to criticize the concept itself or to consider it as a dangerous instrument.

In the Court of Milan's decision the fundamental right to health is re-asserted and is properly directed towards identities of people and unrestrainable freedom of acknowledging one's own identity and clinging to it as regards crucial decisions.

When a fundamental right is asserted, all the consequences which an abstract application lets one imagine are not automatically corroborated; there is a problem of consistency with other rights of the same level, and there is also a problem of plausibility, which must be faced by confronting the given conditions and the requirements within the juridical norms (a matter of measures once again): the conformity with the conduct in compliance with/ or violating norms; the compulsoriness of observance or at least the liability to punishment of transgression; the agreement with ethical guidelines spread within the professional categories concerned. In short, we mean the conditions of effectiveness of the norm, which distinguish the practicable norm from the unrealistic one and which defend the rule from delegitimization involving the ineffective and non-applied norm.

It is important to realise that the most clear and complete assertion of the fundamental right must not be affected by the needs of juridical and factual feasibility.

The nucleus of the fundamental right - or rather, the entanglement of the fundamental rights -freedom-identity and health - is at the origin of the problem of consent. The core itself is expressed more completely by formulating the principle according to which the evaluation of the relationship between the therapeutical suggestion, its consequent risks and benefits, and the possible well-being, is exclusively up to the subject. The principle becomes the unrestrainable and unreplaceable realisation of self-perception and the performance of freedom which sticks to the expression of one's own identity.

The nucleus is accompanied and, so to speak, "attended" by duties and rights of the physician -patient relationship, the definition and discipline of which do not reflect a total, integral and immediate expression of the fundamental right (or rights). Differently, it represents a composition of the latter with equally defended interests and conditions of feasibility in prescribing terms: this is the field of the above-mentioned "measure".

The principle is, anyway, the term of comparison, interpretation and integration of all the concrete prescriptions; there are also moments and aspects of the relationship within which search for composition and measure have no meaning and space. Therefore, in these, the fundamental right reappears as an unquenchable value in all clearness. For example, this is the situation in which an aware patient expresses his/her denial to treatment, or the one in which the medical proceeding prevents the patient from the possibility of judging therapeutical interventions in order to either consent or refuse them-.

In substance, the physician may demand for a plausible measure - in fact, a wider measure than the usual one - as to his commitment to the listening and to the support for decision. He also must be aware that the value of the freedom of the conscious patient is an absolute one, since based on complete consistency between the subject's health and identity. All this means that expressions such as "professional authority" or even "professional power" (this was the term used in the art. 29, 5th paragraph, of the Code of medical deontology 1995; the term is now substituted with "independence" in the Code of 1998) used to demonstrate that the good for the patient - not a possible aspect of his/her own good - is known by the physician since scientifically competent as to distinction between health and illness, well-being and discomfort, and as to individuation of therapy's opportunities and extension of survival, belong to an assumption of the role of the physician and the position of the patient which have no longer place either in our law or in professional ethics[12].

Risks of entanglement.

The close relationship between right to health and the different values of a person must not be misunderstood. One could be tempted to infer from the assumption which states that freedom (article 13 of Const.) is first of all the freedom of disposing of one's own body, and from the further assumption, as regards these choices, that the subject "puts into discussion the idea he/her has of him/herself and of his/her own life's expectations" (Court of Milan), the consequence for which any injury to one of these values, in the patient-relationship - becomes damage to health as a condition of total "eurythmics" of the mind-body unity.

From the two unexceptionable assumptions - the addition of governance of health to the fundamental right to self-determination and the extension of the idea of health to the unity of person meant as a mind-body unity - one could come to the logically groundless conclusion, by ascribing injury to the "goods" - such as freedom or identity -, which are perceived by the subject as features of one's own person, to health as a "good", in its mental and social aspect.
Such an act could certainly ease the compensation for and assessment of damage, but it would make of health a "black hole" in which all values of a person could be subsumed.

A superficial consideration may let think that this risk might have been incurred in one point regarding the grounds for judgment (the one in which a necessary implication between injury to freedom and *"the alteration of the general health condition understood in objectively physical terms"* is mentioned), a risk from which the Court of Milan clearly evades when individuates the responsibility of the defendant expressed in terms of injury to the *"absolute right to liberty (.) ... of self-determination"* from which a remarkable alteration of self-image and self-consideration sufficient for creating an event-damage of a psychic and moral kind derives.

The problem "of social right"

Finally, the difficulty to keep coordination and compatibility between this concept of liberty-health and the one of health as a social right must be mentioned[13].

The compact and objective "health-good" pattern breaks up into a complex series of situations of conflict, considering i.e. the goals of survival, the tolerability of life, subjective well-being and medical evaluation, the efficiency of treatment, and the need for controlling one's own destiny or one's own body by the patient. The evaluations that legislator has to make multiply, and their multiplying depends on the fact that the simple structure of the pattern (covering and meeting a limited area of interests) is replaced by a wide and differentiated range of interests in which a hard composition must be made.

The Constitutional Court had to intervene more than once on the problems concerning norms imposing limits and conditions to the access to health care. The principle of compatibility between meeting the right to health and the needs of collectivity which imply also a reduction of health expenses, has frequently been stated, however, always by protecting the essential and unrestrainable nucleus of the right to health which must have unconditional defence. (see. i.e. const. Court, no.309/99, no.267/98, no. 304/94, no.218/94). The concept of health based on observable standards could be suitable for a drawing with concentric rings which enables to individuate this so-called "unrestrainable nucleus". It seems difficult to be able to do something similar as regards this concept of health which entangles the aspects of inner life, felt and experienced by the subject.

It will be necessary, maybe, to justify a relative impermeability of the concept of health used for social purposes compared to the one for the physician-patient relationship. But the thing does not seem to be very easy.

THE DISCIPLINE OF CONSENT IN THE NEW SOURCES

The whole issue concerning the Court of Milan is based on a perspective of constitutional norms and common principles.

Nonetheless, today there are other sources, a few of which could not be taken into account at the time of decision. We are referring, first of all, to the norms regulating the consent of clinical experimentation . Such an issue, is surely specific and has peculiar aspects as regards the clinical practice and the physician-patient relationship, considering that the proposal has no direct therapeutical value, or has no therapeutical value at all, for the single participant. The issues is also concerned with the different importance and structure of the problem risks-benefits. However, independently of the special character of the discipline, it is liable to represent a normative realization of the common principles one refers to as regards the matter of freedom of disposing of one's own body and of decisions about one's own health, and it is liable to offer, therefore, indications which could be extended by analogy or as realization of common principles as to the common cases of consent to the medical treatment, and with more significance to those cases in which the patient faces less tested or more dangerous decisions.

The *European Convention for the protection of human rights and dignity concerning application of biology and medicine* was agreed on 4th April 1997.

In articles 5-9, Chapter no. II, the principles concerning consent are dictated, starting from the one which declares the necessity for free and informed consent for any intervention within the field of health (Art.5). The necessity for adequate information about the purposes and nature of intervention, consequences and risks are established. Cases of people incapable of consent, of under-aged people and of people with temporary incapacity or mental disorders are established. In a general sense, the necessity for authorization by a representative, an authority or a person determined by the law is provided for. Finally, it is established that one has to consider the previous wills expressed by a person who is not able to express consent at the moment of therapeutical decision (art.9: opening to the living will).

In 1997 the guidelines adopted by the European Agency for the Evaluation of Medicine Products (EMEA) are welcomed in Italy during the plenary session of the permanent Committee for pharmaceuticals for human use [CPMP of 17 h July 1996, (E6 document: Good Clinical Practice: Consolidated Guideline -CPMP /ICH/135/1995)].

The document includes a *definition* of informed consent (glossary, 1.28) which belongs to the normative lexicon. The consent is defined as

> "a procedure (a word to be stressed, editor's note), by means of which a subject voluntarily accepts to take part in a particular clinical study, after having been informed of all the aspects of the study concerning his/her decision. Informed consent is documented by means of a form of written, dated and signed informed consent ".

The necessity for consent is then established among principles in no. 2.9. and 3.3.6, while the problem of consent is regulated by no. 4.8 with complex rules, among which:

- *Integral reference* to the principles of the Declaration of Helsinki (4.8. 1);
- *The forbiddance of coercion* or *"undue influence"* (4.8/3);
- the forbiddance of any *"language " which forces the subject to renounce, even only apparently, his/her rights or exempts, or seems to exempts the experimentator from his/her responsibilities (4.8.4.);*
- *The duty of complete information* (4.8.5.);
- the requisite for *a non-technical, practical language, comprehensible to the subject* (4.8./6);
- the duty *of let people the necessary time to decide and the possibility to be informed,* as well as *of giving satisfying answers to any questions* (4.8.7.);
- the necessity for the signature of the person who *"led the discussion concerning informed consent"* (4.8.8.);
- finally, the *function of the witness* in case of impossibility to read or sign: he must *assist the discussion* about informed consent and attest that what is stated in the form was *accurately explained* to the subject and *apparently understood* by him/her (4.8.9.);
- the analytical forecast of all the things that must be explained to the subject in the discussion, before the form, follows in 4.8. 10.

As seen, the international normative sources and the legislative sources clearly refer to the documents of a deontological kind, especially to the principles of the Declaration of Helsinki.

But deontological sources have importance even without expressed reference, both as elements of extra-normative integration of concepts of professional diligence (and, consequently, of fault) and of general provisions for correctness and good faith, and as interpretative instruments proper to the specification of general principles such as adequacy of information, freedom of consent and, finally, the principle of respect for self-determination.

The history of international deontological documents is long and interesting. We recall the most recent passages of actual practical relevance.

In 1993 the CIOMS (The council for International Organizations of Medical Sciences) approved *the International Ethical Guidelines* as regards biomedical research, of which the early 15 ones concern informed consent. Their characteristic is their being more analytic than any other source, as to the definition of informed consent, to the indication of necessary information and, especially, to the consideration of all the phases and modalities of *the procedure of elaboration and development of consent* (see g. no. 2 and 3); so analytic that they specify which stages are coming first or later, in such a way to foresee, so to speak, the case in points of *unjustified deceit, undue influence, intimidation, undue incentive (g. 4),. the problem of the consent of child is* taken into account, by imposing straightforward information according to his/her intelligence and maturity; respect for the child's denial (g. 5). The problem of the research on individuals affected by behavioural and mental disorders is equally faced (g.6), by considering their possible collaboration to informed consent and the instruments of integration and substitution

of consent, up to the research made on *prisoners* (g.7) or other *socially vulnerable groups* (g. 10) (see the document published in Suppl. no. 4, April 1994 in the magazine "Aggiornamenti sociali", 1994).
The *"Guida europea di etica medica"* (later modified with the title *"Principi di etica medica europea",* Parigi, 1982) establishes the reference for all deontological codes of the European Community.

The problem of consent is dealt with in art. 4 by giving special attention to the problem of information compared to the risks and the benefits of intervention. The "Guide" itself inserts at a deontological level the principle relevant to the purposes of evaluation of the costs-benefits-relationship, according to which

> "the physician cannot substitute his/her own conception of quality of life with that of his/her patient ".

Consistency with the foundations of the Court of Milan's decision is evident.

The Italian experience draws attention on two national documents. The document of the *National Bioethics Committee* (20th June 1992), devoted to the "information and consent to the medical act", defines consent as "the legitimisation and the basis of medical act" and as an instrument for establishing what the ethical language calls the *"therapeutical ally"* between the physician and the patient.

In the conclusions and proposals, the Committee formulates then a wide prescriptive guideline about he necessary *behaviour* to the acquisition of consent and about the *condition of efficacy* of the latter, which correspond to the four conditions referred to by the Ministerial decree on the Ethics Committees, but that the document ascribes to *general consent.*

To limit ourselves to a note, one should stress that those solutions, which the ethical perspective makes more accessible, appear in this document as in other ones. Also the juridical language is learning to welcome them, - solutions concerning the importance and the treatment of the condition of limited capability, and concerning the roles of persons close to the patient, the relevance of which is emphasized compared to the one of the eventual legal representatives.

The 1998 FNOM's *deontological code* has improved the prescriptions concerning informed consent, already enriched in the previous one (1995). The respect for the patient's will does not appear specifically in the oath yet, but it becomes an entry, under the name of respect for *freedom*, among the duties and the general tasks of the physician (art. 5, paragraph 1); the respect for the fundamental rights of the person is the first norm which regulates the relationship with the *"citizen"* (expression which has substituted that of “patient” in this context, art. 17). A complete head, the IV of title III, is devoted to information and consent of the patient. Here the norms on the information of the patient (art. 30 e 31) set off, in which not only the contents of information but also the way to furnish it are prescribed: the attention to the capability of understanding, on the one hand, the caution for the cases in which information may

> "cause worry and pain to the person " (art. 30),

on the other. One should stress also the provision for

> "the documented refusal of the person able of understanding and will"

the observance of which is imposed to the physician

> "since no treatment is allowed without the will of the person";

and further more, the duty to

> "keep to the will for treatment, freely expressed by the person"

is confirmed in art. 34, according to which

> "the physician, if the patient is not able to express his/her own will, in case of serious danger for life, must consider what was previously expressed by the latter"

(opening to the *living will);* even in the case of under-aged people or of insane adults, except for the rights of the legal representatives, the physician must

> "take into account his/her will"

and

> "give information (...) compatibly with age and with the capability of understanding".

To crown all the instructions, paragraph 2 of article 32 (regulating the acquisition of consent) must be stressed. It states that consent, expressed in written form in those cases foreseen by the law, or where an unequivocal manifestation of will is necessary,

> "is integrative and not substitutive for the informative process as per art. 30".

INFORMED CONSENT: ACT AND PROCESS

I would like to stress one point which bestows significance on the development of the topic, the richness of which is still to be discovered. Here, the shifting of the focus of the normative attention is meant. This suggests a new perspective of the theorical observation on the problem of consent - or rather, the consideration of the informed consent not only as a category of the action but also as a *"process"*.

The act of consent obviously exists - the moment on which an actual will expresses itself and is opportunely conveyed and documented. However, this consent must be considered not only as the verge of an itinerary, which can take on the connotation of a procedure from a juridical point of view, but also a factual point of view - stages following one another which must be foreseen coordinated and performed carefully.

To look at the process of development of consent means to guarantee and evaluate the condition of process, not or not only the act concluding and representing the climax of it. Therefore, one has to consider the mode of interaction, material conditions and training of the staff as well as information of people around him/her; to consider the time devoted to information and the time devoted to the listening to the patient, as conditions which are as necessary as the contents and clearness of information. One has to consider the way a decision is conveyed by turning one's eyes not simply to the act, but also to the process, and, thus, by adopting a hermeneutical knowledge able to appreciate the whole language of the person from the words to the attitudes as the discipline of law has already done in some cases, (i.e. interruption of an incapable woman's pregnancy.). Only by means of consideration and of a discipline of the process, it is possible to stress elements such as weakness and power of consent, its resistance or frailty; and to give the right importance to the confusion and anguish co-existing, as everybody knows, with the use of mind.

One has to consider with the greatest respect those prescriptions involving the manifestation of consent, and especially the creation and the subscription of the form of informed consent, understood as the conclusion of the process of decision-making, which the form has the purpose to attest and not to substitute. The form is a means of sure documentation concerning something which took place a point in time, lingering on the necessary stages. The act of consent itself is not enough.

This does not mean to go beyond the contractual perspective but, contrarily, to increase the value of it.

In the contractual matter itself resides the rule which manages the process of consent, with its guarantees sufficient for the patient as well as for the medical or paramedical staff, and for the medical structure interacting with him/her. We are coming back to the general provision for correctness mentioned in the first section of the report, as an instrument giving juridical, and not ethical, measure to the physician's duty and to the patient's needs within the health-relationship. The general provision for correctness and good-faith in the development of the negotiations, interpretation and execution of contract represents a principle of the development of the production and performance process of the contractual decision. The principle is capable, if adjusted to the needs of the discussed field, to indicate a measure for defining the behaviour requested for the efficacy of the process and the conditions of respect for the person and for the freedom of the one who decides as well as the conditions of guarantee for the subjects supporting the decision: without fostering illusions on the possibility, according to the legal rule, of going beyond an honest discipline on social behaviour, which necessarily leaves to other prescriptions and other languages the orientation to the inner attitudes.

Paolo Zatti, Professor of Law, Faculty of Political Sciences, University of Padua, Italy.

NOTES

[1] Modugno F. (1995), *I "nuovi diritti" nella giurisprudenza costituzionale,* Torino; Cocconi M. (1998), *Il diritto alla tutela della salute,* Padova.

[2] About the constitutional fundament of the freedom of disposing one's own health and for a picture of doctrine and jurisprudence, see, D'Arrigo C. M. (1999), *Autonomia privata e integrità fisica,* Milano.

[3] About this point, see Berlinguer G. (1997), *Etica della salute,* 2nd ed., Milano, p. 19.

[4] *Ibidem,* p.20.

[5] About the international discipline of the right of health see Sinagra's, Cruciatti's and Scalabrino-Spadea's (1990) contributions in *Le médicin face aux droits de l'homme,* Padova.

[6] D'Amico P. (1992), *Il danno da emozioni,* Milano; Ziviz (1999), *La tutela risarcitoria della persona-danno morale e danno esistenziale,* Milano.

[7] Gadamer H.G. , *Dove si nasconde la salute,* 1st italian edition, Milano, especially 28, 103, 120, 135.

[8] Cocconi M. (1998), *Il diritto alla salute,Padova,* 98.

[9] Gadamer H.G., ibidem, 138.

[10] Levine S. (1997), *Guarire per vivere*

[11] Zatti P. (1997), *Il principio di correttezza nel rapporto medico - paziente,* in atti del Convegno "*Il consenso informato*", Centro Nazionale di prevenzione e di Difesa sociale - Fondazione dell'Istituto Europeo di Oncologia.

[12] Power is a tenacious word in medical language, see, Iadecola (1998), *Potestà di curare e consenso del paziente,* Padova). The reference of the principle of "integrity and independence", is of a different kind, see Cattorini (1992), *I principi dell'etica medica,* in *Trattato di Bioetica,* edited by Bellino, F., Bari; on the idea of authority in medicine see Gadamer again, *Dove si nasconde la salute,* ibidem, 127.

[13] Dell'Erba, Fineschi (1998), *La tutela della salute,* Milano; Gallo – Pezzini (eds) (1998), *Profili attuali del diritto alla salute,* Milano; De Carolis (1999), *Potere di scelta e tutela dell'utente del settore sanitario,* in *Sanità Pubblica,* 783.

PAOLO VINEIS

THE TENSION BETWEEN ETHICS AND EVIDENCE-BASED MEDICINE

THE TENSION BETWEEN ETHICS AND EVIDENCE-BASED MEDICINE

I believe that the most intriguing problem of modern medicine is the tension between population-based knowledge and individual-based decision-making. Both clinical decisions and resource allocation are based on inferences derived from the study of populations: investigations of clinical epidemiology in the case of clinical decisions, and several different sources of information (including technology assessment and economical evaluation) in the case of resource allocation. The reference to population-based studies means that choices are targeted to the average individual, with an average clinical response and average needs. The beauty and the difficulties of medicine are related to the transferral of population-based knowledge to the single individual. The good physician is the one who is able to achieve the best trade-off between the available resources, the standards of practice (founded on evidence) and the personality and needs of the single patients, in an intertwining of objective and subjective elements. According to Muir Gray[1], the three "eternal truths" in health care are "evidence, values and resources", again expressing a tension between individual vs. collective aspects.

EVIDENCE-BASED MEDICINE

For a long time medicine was based on sporadic observations of patients, often characterized by the lack of formal sampling techniques, biased collection of information and an understatement of measurement error. Randomized controlled trials were introduced in the 40's with the pioneering work of Sir Bradford Hill, and represent the correct technical tool to achieve evidence on the efficacy of treatments.

More recently, Evidence-based medicine (EBM) has become popular amongst clinicians and epidemiologists as a tool to facilitate the translation of scientific research into clinical practice. The principles of Evidence-Based Medicine (EBM) include: the use of quantitative estimates of efficacy, including confidence intervals; the reference to Randomized Controlled Trials (RCTs) as the gold standard of medical evidence; the conduction of systematic reviews of the evidence, usually based on meta-analyses (e.g. the Cochrane Library)[2],[3]; the use of scores to assess the qualitative level of the studies; the search for publication bias (use of funnel plots and similar approaches) and other sources of bias; and systematic and rational approaches to transfer of research into practice (EB-guidelines).

C. Viafora (Ed.), Clinical Bioethics. A Search for the Foundations, 131-137.

Basically, EBM consists (a) in the conduction of systematic reviews of the evidence; (b) in setting rules for the use of evidence in clinical decisions, for example through the elaboration of guidelines for clinical practice. It should be clear that (a) and (b) are two distinct steps: (a) is principally based on scientific principles for causal assessment, while (b) involves extra-scientific elements, such as reference to values. Typically, step (a) consists in a meta-analysis of the available randomized clinical trials, while step (b) leads to recommendations. Not necessarily there is a strict link between the two. As the guidelines of the American Society for Clinical Oncology suggest, a high level of scientific evidence can be followed be a low-level recommendation (for instance because the drug is expensive, or not tolerated by the patients), and viceversa. Thus, a tension between evidence and values is acknowledged when EBM is used to develop guidelines for clinical practice.

It is my opinion, in fact, that the distinction between assessment of the evidence and decision-making is essential. Although decision-making should be clearly based on a degree-of-evidence principle, action can be taken even with a low degree of evidence, depending on the circumstances. To give an example, drugs have been banned even if there was a weak suspicion - based on case reports - of a toxic effect (i.e. on the basis of poor evidence), because their benefit was modest or absent: this is the case of gangliosides and the risk of Guillain-Barré syndrome. The important principle underlying EBM is to summarize all the available evidence, to create awareness about the areas in which evidence is lacking, and to use the best evidence for decision-making[4].

INDIVIDUAL VS. POPULATION IMPACT

Most systematic reviews of the literature use the Odds Ratio as a measure of effectiveness (see for example the already mentioned meta-analyses contained in the Cochrane Library). However, a simple relative measure of effectiveness may be misleading: although it is appropriate for causal assessment, it is not to estimate the overall impact of clinical practice.

Let us imagine that an intervention reduces mortality by 20% (i.e. the Odds Ratio is 0.8). If the absolute risk of death in the population is 1%, then the absolute reduction of risk is 0.002 (that is, 1 % multiplied by 20%): the intervention will be able to avoid 2 deaths per thousand persons treated. If the death rate is 30%, then the deaths avoided will be 30% multiplied by 20%, i.e. 6%, meaning that we will avoid 6 deaths every 100 persons submitted to the intervention. The inverse of the prevented deaths is called the Number Needed to Treat (NNT), which is 110.002 (=500) in the first example and 110.06 (=17) in the second example: In other words, we need to treat 500 subjects to avoid one death in the first case, vs. only 17 in the second case. Clearly, the NNT conveys much more information than the Odds Ratio, since it includes the absolute frequency of the event (in this case the death rate). Although the relative effectiveness of the two interventions is the same, its efficiency - related to the best use of resources - is much greater in the second case. However, for the single individual this reasoning is not straightforward, since he will always have a 20% reduction of his personal death rate.

Along the same lines, Geoffrey Rose, in an important book called *The Strategy of Preventive Medicine*[5], has stressed that we should expect greater gains, in preventive policies, from strategies aimed at large sectors of the population rather than at small, high-risk groups. Rose's idea is simple and typically utilitarian on moral grounds: the vast majority of the cases of disease do not occur in the minority of subjects who have extreme risks, but in those with average or low risks. This means that the most efficient intervention is one that slightly decreases the individual risk but considerably greater efficacy in reducing diminishes the population burden of disease (low relative risks, high attributable risks). Table 1 shows how different measures of efficacy give different pictures of the benefits of two clinical practices: the relative risk reduction suggests that mammography has a mortality from breast cancer, while the NNT is much more favourable for cardiac rehabilitation (31 subjects needed to treat vs. 1592).

Table 1

	MAMMOGRAPHY	CARDIAC REHABILITATION
RRR	34%	20%
ABSOLUTE RISK REDUCTION	0.06%	3%
%EPISODE-FREE PATIENTS	99.8 vs. 99.82%	84 vs. 87%
NNT	1592	31

Table 2 shows that different clinical practices have radically different NNTs (i.e. population impacts), and also that thay have different NNTs depending on the characteristics of the populations that are studied.

Table 2.NNT for different types of clinical practice

CAROTID ENDARTERECTOMY	9
SERIOUS HYPERTENSION	15
MODERATE HYPERTENSION	700
MODERATE HYPERTENSION OVER 60	18

TREATMENT OF ANGINA	25
ASPIRINE IN ASYMPTOMATIC AMERICAN PHYSICIANS	500

The conflict between individual and population is multifaceted, as expressed by Muir Gray[6] in the following table:

Table 3. From Muir Gray, ref. 1, modified

Clinical care	**Public health**
For individuals	For populations
Treatment for those who feel ill	Treatment for those who feel well
Low NNT	High NNT
Decisions unique to the individual	Decisions common to populations
Difficult to produce systems and guidelines	Easy to produce Systems and Guidelines

Muir Gray's book on Evidence-based Health Care[7] is an attempt to stress the need for scientific evidence in order to macompetent and effective decisions in health services. The general idea of the book is that the best health care is the one that offers effective clinical practice in the most efficient way to the patients who can benefit mostly.

But who are the subjects who can benefit mostly? And who are those who will not benefit at all? We will touch upon this problem in the next paragraphs.

ETHICS, EBM AND MANAGED CARE

If we consider the Mission Statement of the Cochrane Collaboration, a reference to ethics is clearly present, with a particular emphasis on informed consent:

> "The Cochrane Collaboration is an international organisation that aims to help people make well-informed decisions about healthcare by preparing, maintaining and promoting the accessibility of systematic reviews of the effects of healthcare interventions. Principles. The Cochrane Collaboration's work is based on nine key principles: Collaboration, by internally and externally fostering good communications, open decision-making and teamwork. Building on the enthusiasm of individuals, by

> involving and supporting people of different skills and backgrounds. Avoiding duplication, by good management and co-ordination to maximise economy of effort. Minimising bias, through a variety of approaches such as scientific rigour, ensuring broad articipation, and avoiding conflicts of interest. Keeping up to date, by a commitment to ensure that Cochrane Reviews are maintained through identification and incorporation of new evidence. Striving for relevance, by promoting the assessment of healthcare interventions using outcomes that matter to people making choices in health care. Promoting access, by wide dissemination of the outputs of the Collaboration, taking advantage of strategic alliances, and by promoting appropriate prices, content and media to meet the needs of users worldwide. Ensuring quality, by being open and responsive to criticism, applying advances in methodology, and developing systems for quality improvement. Continuity, by ensuring that responsibility for reviews, editorial processes and key functions is maintained and renewed".[8]

Therefore, a strong ethical commitment is made in the very premises of EBM activities, with emphasis on transparency, scientific rigour and the rights of users.

However, the picture is different if we consider uses of EBM in the context of "managed care", a slightly different perspective. Managed care is an example of a "third party" trying to control the use of medical and health care services by influencing the decision making of the patient as well the decisions about the reimbursement of care. This system tends to have profound influence on the professional autonomy of the physician and the decisional autonomy of the patient. Managed care tends to refer to EBM to justify evidence-based choices. However, several issues potentially relevant to ethics can be raised. One important issue is whether evidence is equally available for all relevant problems touched upon by managed care. This is clearly not true: while a large number of RCTs are available on pharmacological treatments (e.g. management of myocardial infarction), emergency medicine and in general organic problems, the amount and quality of evidence concerning other levels of health care (psychological support, deprivation, social networks) is much lower. This is so for objective reasons (these are more difficult issues that are not necessarily prone to an RCT approach), subjective reasons (less interest by scientists) and market-related reasons (less or no investments by the industry).

In addition to the goals of research, also different subcategories of the population can be differently affected by evidence-based managed care. For example, one of the principles of managed care is to base decisions on indicators of cost-effectiveness such as "life-years" gained. However, the elderly can be disadvantaged by this approach, since they can take relatively less advantage of treatments that prolong life. Also, there is evidence that clinical trials are usually conducted preferentially in younger patients. The major trials on hyperlipidemia (4S, CARE, LIPID) have been conducted in patients with 60-75 years of age, but in WOSCOPS the limit of age was 65 years.

SPECIAL POPULATIONS AND THE PROBLEM OF ACCESS

Access to health care has been defined as the

> "timely use of affordable personal health services to achieve the best possible health outcomes"[9]

Special populations can have special access problems, for example the elderly or the deprived. There are several ways in which the elderly can be affected by problems of access to clinical services: for example, in the case of cancer they often underestimate their risk, and have an average of three or more chronic medical conditions, a situation that tends to relax the clinician's attitude towards cancer detection and to hamper screening activities[10]. Also social class is a very potent determinant of decreased access. Hazard ratios for survival may be as much as 60% lower for breast cancer in lower classes compared with their more economically advantaged counterparts. African-american ethnicity was associated with having later disease stages at diagnosis, and this effect was almost entirely explained by social class[11].

IS THERE EVIDENCE IN FAVOUR OF SOCIAL SUPPORT?

In spite of the potential conflicts between ethics and EBM, providing scientific evidence in favour of social support and non-pharmacological interventions is an important goal and a way to reconcile EBM with ethical issues. One example is represented by social support for high-risk mothers. Babies born in socio-economic disadvantage are likely to be at higher risk of injury, abuse and neglect, and to have health problems in infancy. The objective of a Cochrane review[12] was to assess the effects of programs offering additional home-based support for women who have recently given birth and who are socially disadvantaged, compared to usual care. Eleven studies, involving 2992 families, were included. There was a trend towards reduced child injury rates with additional support, although this was not statistically significant (odds ratio 0.74, 95% confidence interval 0.54 to 1.03). There appeared to be no difference for child abuse and neglect (odds ratio 1.12, 95% confidence interval 0.80 to 1.57), although differential surveillance between visited and non-visited families was an important methodological consideration. Babies in the additional support groups were more likely to have complete immunizations.

CONCLUSIONS

EBM is based on strong ethical premises concerning transparency and the rights of the patients. However, there are also potential dangers in EBM from an ethical perspective: there are difficulties in finding a balanced trade-off between population-based knowledge and the needs of the individual patient; too much emphasis is put on RCTs as the gold standard, an appropriate view if we consider pharmacological treatments, not if we are preoccupied with interventions such as social or psychological support (although some randomized trials are available also for these issues, one wonders whether the RCT is the most appropriate tool); the use of EBM for managed care can introduce inequalities related to the lack of relevant evidence for special categories of the population, such as the elderly. Clearly, all these issues require particular attention in the context of changing European health care systems.

Paolo Vineis, Professor of Epidemiology and Biostatistics, University of Turin, Italy.

NOTES

[1] Gray JA (1997), *Evidence Based Health Care. How to make health policy and management decisions,* Edinburgh: Churchill Livingstone.

[2] Blettner M., Sauerbrei W., Schlehofer B., Scheuchenpflug T. (1999), Friedenreich C., *Traditional reviews, meta-analyses and pooled analysis in epidemiology*, Int. J. Epidemiol,28, pp.1-9.

[3] Chalmers I. (1993), *The Cochrane Collaboration: preparing, maintaining, and disseminating systematic reviews of the effects of health care,* Ann NY Acad Sci; 703: 156 Muir –163.

[4] Chalmers I. (1993), *The Cochrane Collaboration: preparing, maintaining, and disseminating systematic reviews of the effects of health care,* Ann NY Acad Sci; 703: 156 Muir –163.

[5] Rose G. (1992), *The strategy of preventive medicine,* Oxford University Press, Oxford

[6] Gray JA (1997), *Evidence Based Health Care. How to make health policy and management decisions,* Edinburgh: Churchill Livingstone.

[7] Gray JA (1997), *Evidence Based Health Care. How to make health policy and management decisions,* Edinburgh: Churchill Livingstone.

[8] The Cochrane Library. Update Software, Issue 1, 2000.

[9] Mandelblatt JS, Yabroff KR, Kerner JF (1999), *Equitable access to cancer services,* Cancer ,86, pp.2378-2390.

[10] Mandelblatt JS, Yabroff KR, Kerner JF (1999), *Equitable access to cancer services,* Cancer ,86, pp.2378-2390.

[11] Mandelblatt JS, Yabroff KR, Kerner JF (1999), *Equitable access to cancer services,* Cancer ,86, pp.2378-2390.

[12] The Cochrane Library. Update Software, Issue 1, 2000.

H. JOCHEMSEN, J. HOOGLAND, J. POLDER

MAINTAINING INTEGRITY IN TIMES OF SCARSE RESOURCES

INTRODUCTION

In many industrialised countries health care expenditures have increased to such an extent, that governments have felt the need to control the cost increase. At the same time the medical possibilities have increased constantly and so has the demand, both quantitatively and qualitatively, in the population for health care provisions. Thus,many industrialised countries governments see themselves confronted with a tension between the cost and the quality of health care. In a number of countries changes in the health care system have already been introduced or are being introduced to reduce this tension (e.g. the Netherlands, Great Britain, Germany and some central European countries).

Costs in health care to a large extent result from actions and decisions of physicians. Changes in the organisation and financing of health care in order to control expenses, tend to affect the position of physicians, in particular their freedom of action. However, a physician's freedom of action entails an important ethical value, and policies that affect it may have significant ethical implications. This leads to the question as to how far it is ethically justifiable to limit the freedom of action of physicians by policy measures, or better, governance instruments[1]. for cost control. In order to answer this question we have approached the issue from two angles.

First, we have made a philosophical analysis of the character of medical practice in order to ascertain the ethical value of the physician's freedom of action[2].

Secondly, we have studied the effects of a few governance instruments as elements of a broader health care system on the freedom of the physicians. In the final part of the study these effects are evaluated in the light of the results of the first part of the study.

ANALYSIS OF NORMATIVE PRACTICES

Our philosophical analysis of medical practice has resulted in the formulation of a model for this practice. This model is built up from the following elements.

a) Medical practice is understood as a practice in the sense of a coherent form of human activity and accompanying competences. In short: practice is competent performance. These competences are grounded in rules. In this context the concept of `rule' does not refer to rules in the sense of 'knowing that', which implies the ability to formulate the applied rules. Rather, it refers to rules in the sense of `knowing how', which is an intuitive awareness of rules, consisting in the ability to

C. Viafora (Ed.), Clinical Bioethics. A Search for the Foundations, 139-152.

act according to a rule and to evaluate the correctness of this application. To clarify this: one can easily see that performing a practice, e.g. playing the violin or practicing medicine, cannot be learned just by theoretical instruction about that practice, but that engaging in that practice is indispensible. Thus, a competence exists in the ability to act according to the (usually implicit) rules of the particular practice. These rules have an intrinsic normative nature in the sense that they apply to a specific practice and constitute the possibility to evaluate the correctness of the actions performed within that practice.

b) In the second place we derived an important notion of practice from the definition of MacIntyre:

> "By a `practice I am going to mean any coherent and complex form of socially established cooperative human activity through which goods internal to that form of activity are realised in the course of trying to achieve those standards of excellence which are appropriate to, and partially definitive of, that form of activity, with the result that human powers to achieve excellence, and human conceptions of the ends and goods involved, are systematically extended"[3]

The term 'internal goods, mentioned in this definition, is often interpreted in terms of goals. But goals are always related to individual or collective actors. And the goals set by those actors do not necessarily constitute the 'internal goods' of a practice. For example, a person could frequently play billiards with the goal of becoming the best billiard's player in an establishment. Yet, one cannot consider this goal as the 'internal good' of the game of billiards. Its internal good is to play the best possible shots, within the rules of the game. Even winning or losing a game is not the internal good, though this will often be a goal of the player. To avoid the ambiguity of the term `internal good' we prefer to speak about the finality of a practice[4]. A practice's finality determines what goals are appropriate within that practice and it contains standards which are decisive for the kind of performance that is adequate within a practice. Thus the finality of a practice leaves space for a number of subjective goals which could be set within a specific practice.

c) Combining a) and b) we describe a practice as a coherent form of human activity in which (mostly implicit) rules, that relate to the internal nature and finality of the practice, define the competences and the standards of adequate performance of that practice.

We combine the different characteristics of such rules by using the term 'constitutive rules'. This concept expresses the view that the normative structure of a practice must be considered as the 'playing field' for concrete goals and actions within that practice. In our view the 'constitutive rules' make a practice recognizable as a specific practice and determine its finality. We call this side of a practice its constitutive side.

To every practice, being a coherent form of human activity, a number of aspects can be distinguished. In this context `aspect' can be understood as an irreducible mode of human experience that also constitutes a way of evaluating the performances within a particular practice. The way a person performs a practice, for instance playing the violin, can be evaluated, among others, from a logical-analytical, a social, an economic, a juridical, an aesthetical and an ethical point of view. The evaluation from these points of view requires rules as criteria. These are provided by the previously mentioned constitutive rules. So, each of the aspects in which practices function provides constitutive rules (mainly of the `knowing how'-type) that define an adequate performance of the practices.

All practices function in all aspects, but the aspect-related rules do not apply to all practices in the same way. A distinction must be introduced here, namely between constitutive rules in a typical (or qualifying) sense and in a non-typical (or foundational) sense. One of the aspects that can be distinguished to every practice, including the corresponding rules for assessing adequate performance, gives a particular practice its own typical character. We call this the typical or qualifying aspect that is directly related to the finality of the particular practice. The rules related to this aspect are the constitutive rules in a typical or qualifying sense. The rules related to the other aspects are constitutive in a non-typical, foundational sense. The functioning of the non-typical rules is guided by the qualifying rules. An adequate performance requires the *simultaneous realisation of the rules* (that in the assessment function as *norms*) related to the various aspects.

Before we explain this abstract analysis by applying it to medical practice one more element should be added. The constitutive side of a practice concerns the normative constitutive rules that relate to the various perspectives (aspects) from which a performance of a practice can be assessed. However, such an assessment always involves a specific *interpretation* of the rules. (Cf. the interpretation of a piece of music depends also on the ideas of the performer on the way the piece should be understood and performed). In other words, the performance of a practice is always from a wider interpretative framework on the meaning of that practice for human life and for society and, hence, on the *direction* performances of that practice should have. We call this the regulative side of practices. At this level world views have a regulating function: depending on their view on the meaning and the coherence of reality, people act differently in concrete practices. The (constitutive) structure does not determine the direction of the performance; the rules of a play do not determine the course of an actual play but only which courses are correct. It is part of the character of normative practices that they can only be `opened up' by regulative ideas about the meaning and structural coherence of human experience. Put more generally, human behaviour, and also (theoretical) reflection on it, is regulated by world views. This also applies to ideas about the adequate performance of medical practice.

STRUCTURAL ANALYSIS OF MEDICAL PRACTICE

Constitutive rules

What does this theoretical framework of social practices mean for medical practice? To trace the constitutive rules of medical practice, we start from what the Dutch physician/ethicist prof. G.A. Lindeboom called `the core medical situation':

> "the core medical situation is where a sick person summons a physician for help. The sick person is a human being in need, because of his physical or mental condition. In his need he asks for help of someone, whom he believes, is able and willing to give it"[5]

This `core medical situation' consists of three elements: 1) the appeal of the suffering patient, his complaint; 2) the special competence of the physician; and 3) the professional character of medical practice.
The latter element is basic for the former two. A profession can be described as a `body of persons engaged in a calling' in which the calling can be formulated as `the rendering of a public service'[6]. Some authors speak of the selfless rendering of a public service[7]. The oath or promise that professionals swear before they begin to practice their profession should be seen in this context. This oath or promise has to make clear to the (potential) patient/client that the professional will use his specialised knowledge and skills in the best interest of the patient/client[8]. This guarantee for the trustworthiness of the professionals is required because they deal with vital interests of their patients/clients (the first element) who cannot themselves control whether the service rendered is in their best interests because of the specialised character of the professional activities (cf. element 2 above). To maintain the confidence of the population the profession clearly needs to control the quality of the service rendered by the individual members and to be willing to render account of their activities and policy to society at large.

Taking together the three elements of the core medical situation we conclude that the physician/patient relationship is essentially a relationship of assistance and care. In our view the principle of care - or to be more specific: the principle of benevolence[9] - indicates the `meaningkernel' of the ethical aspect. So, in terms of the analysis presented in the former section we can say that the rules related to the ethical aspect are constitutive in a typical sense for medical practice; in other words, medical practice is ethically qualified. This also means that the ethical way of assessing medical practice regulates the ways of assessment corresponding with the other constitutive rules. An example may clarify this.

In our society a physician earns a living by his practice. Yet no one will claim that the specific character (MacIntyre's `internal good') of medical practice consists of making money. Neither will anyone evaluate the quality of a physician's work in amounts of money. This is because the finality of medical practice is determined by the ethical (or moral) aspect of which the meaning-kernel is benevolence: helping people who are in trouble due to physical or mental suffering. Yet the economic aspect is inherent and important to medical practice. For a physician earning money is not just a minor detail. Furthermore, making an efficient use of his time and of the available resources is part of a competent performance of medical practice. However, what is understood by economic waste within medical practice should be determined by checking what is necessary and effective from a medical-ethical point of view, since the finality of medical practice is qualified by the ethical principle of benevolence. This also means that policy instruments for cost control should be evaluated primarily in the light of the principle of benevolence and not primarily on the basis of their effects on the income of physicians or on the freedom of treatment of the physician in itself. On the other hand it is contrary to the constitutive elements of medical practice to demand from the physician to select patients for certain treatments on other than medical grounds. The physician has to do everything which lies in his capacity to help and assist the patient.

So, the rules related to the economic aspect are constitutive for medical practice but in a non-typical sense. The same can be argued for the other aspects e.g. the juridical, the scientific and the technical aspects. We will elaborate briefly on this latter aspect.

In trying to help a patient a physician should apply his scientific knowledge and technical skills. Modern specialised medicine has a highly technical character and is unthinkable without technology. In many situations technology may even seem to be constitutive for medical practice in a typical sense. This, however, is incorrect and even dangerous. Technical possibilities should not be used just because the technique is available and could have a physiological effect. Whether an available technique should be used or not depends on the question whether in the light of the principle of benevolence the technique is medically - which is broader than physiologically - indicated or not. That medicine is not technically but ethically qualified is also demonstrated by the fact that, even when it is technically impossible for a physician to cure, it is still his task to care for the patient.

Yet, medical technique is clearly constitutive for medicine in a non-typical sense. Not only is medicine impossible without technology, but the medical-technical aspect also `colours' the care that a physician should give. It is evident, that the care of a physician for a patient is of a completely different character than the care of a mother for her child, or of a geriatric helper for a senior citizen. The care of a physician for his patient gets its character from the field of competence of the physician. The care relates to a patient's ill-being (a disruption in the biotic or mental aspect of normal life). A physician defines and treats diseases according the skills and knowledge he gained during his education. So, the way of looking at disease, healing and health within medical education and medical science is included in the constitutive rules of medical practice. (Figure 1 gives a schematic representation of the main elements of medical practice).

Fig. 1. Schematic representation of the main elements of medical practice

CORE MEDICAL SITUATION : sick person asks physician for help

three elements:
1. Complaint of patient
2. Competency of physician
3. Professional character ~ oath or promise

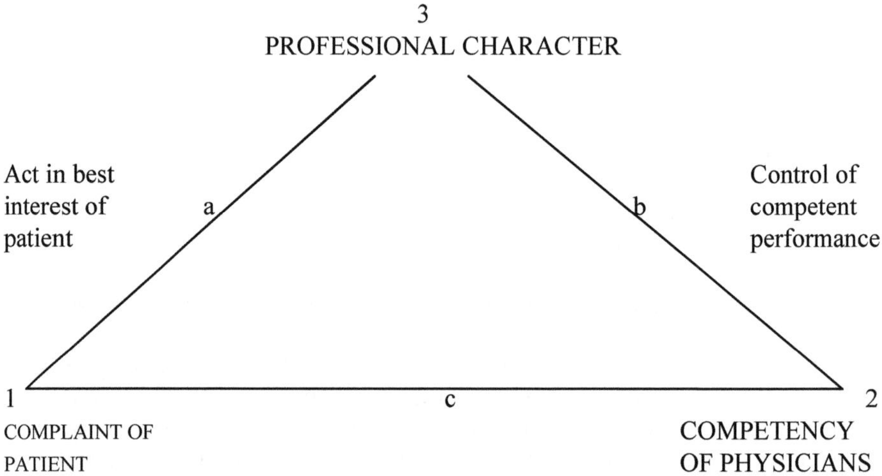

a: qualifying constitutive rules; meaning kernel: benevolence
medical ethical codes
b: constitutive rules in non-typical sense: medical-technical, economical, juridical, etc.
c: medical-professional standard

This model sustains the traditional view that medical treatment requires a medical-professional indication. We conclude, without further arguing the case, that medical treatment must be positively justified by a medical indication. The request or consent of a (competent) patient is an essential precondition for treatment but not its ground. (In this context the constitutive rules related to the juridical aspect play a role. The available space does not allow a further elaboration of this).

REGULATIVE RULES

We saw in the previous paragraph that normative practices, like medical practice, not only have a constitutive or structural side but also a regulative side. This pertains to view of life and world view that provide an interpretative framework for human experience and action. In medical practice regulative ideas of both physicians and patients about health, sickness, medicine, the good life, etc., will influence the way in which the constitutive rules are applied.

Often the regulative ideas or conceptions in medical practice remain quite implicit. Currently, ideas that transcend the positive knowledge of the sciences and humanities have lost their legitimacy in public debate. World views and religious beliefs are usually seen today as subjective perspectives which cannot be rationally justified. But the fact that it is impossible to give a scientific justification of our more fundamental beliefs and convictions does not mean that they are unimportant. In some respects one could say that the rupture between the classical and the modern view on medicine is a rupture between medicine as embedded in a metaphysical or religious world view and modern medicine that has lost its embeddedness in this sense. But just when medical practice is no longer regulated by a common world view, it will be threatened by unrealistic expectations. In the modern age only scientific reasoning seems to be obligatory. Science has become its own regulative framework. This is the reason why science and technology could play a seemingly autonomous role in medical practice. To maintain a sustainable health care system it is important to make predominant regulative ideas explicit and discuss them in society. A realistic view on what medicine, both from a medical-technical and from an economic point of view, can do for the human condition is required to maintain a health care system that is not only sustainable but also ethically justifiable.

This analysis of medical practice leads us to the formulation of the following *general principle* regarding health care reforms:

In a policy of cost containment in health care preferably those governance instruments are used that are in keeping with the internal finality of medical practice. This pertains to instruments that

- respect the irreducible responsibility of the physician to serve primarily the interests of the patient and
- challenge the physician to an optimal performance of his competence (in the full diversity of rules and principles that apply to medical practice) and
- respect the freedom of treatment the physician needs to realise such a performance.

The physician in the health care system

Having formulated a normative analysis of medical practice we will now investigate what this should imply for the health care system.

The health care system gives the organisational framework in which health care delivery takes place. The health care system lays down the authorities and responsibilities of persons and bodies involved in health care and supplies rules for their mutual relationships. In this way, the health care system has a co-ordinating function in the tuning of all kinds of individual and collective activities of parties in the care field. Five parties can be distinguished: 1) patients; 2) care workers; 3) care institutions; 4) suppliers of equipment and resources; 5) health insurance companies. Government finds itself outside the actual playing field and from this position gives rules about how the game must be played[10].

In our structural analysis we have called medical practice a normative practice. The reason is that our analysis led us to the conclusion that the structure of medical practice can be characterised by (mostly implicit) constitutive normative rules, or norms. These norms are related to the various aspects and they provide criteria for assessing whether a particular performance of medical practice is adequate. An adequate performance of medical practice requires the simultaneous realisation of the various norms, in which the qualifying ethical principle of benevolence regulates the realisation of the other non-typical norms that are valid for medical practice.

In this view the health care system, that provides the organisational framework for the realisation of medical practice (among others), should be subservient to the simultaneous realisation of the various norms that should be observed in medical practice. Again, this realisation should be regulated by the principle of benevolence.

What does this mean for the health care system? To investigate this we briefly describe three different kinds of health care systems as ideal types (see Figure 2)[11] .

Fig. 2. Three theoretical, ideal-type health care systems

	centrally planned health care system	`guild-free-choice' system	managed competition
goals/ends	optimal medical care and a fair distribution among the population	optimal medical care and maintenance of a trust relationship between physician and patient	optimal medical care and an efficient allocation of (medical) resources
level of decision-making	macro-level (government or local authorities)	meso-level (profession and other groups)	micro-level
property rights	public	private (agreement)	private (contract)
information dissemination	vertical through hierarchical imperatives	asymmetric distributed, imperfect dissemination	information embodied in prices, sponsor can publish information
coordination	budget mechanism (imperatives)	free choice of physician (prescription and treatment)	market with aid of a sponsor who manages the competition
main governance problem of system itself	loss of quality inefficiency	high and uncontrolled cost increase	lack of solidarity no basis for trust
other governance problems due to a malfunctioning of the systems	rise of `upper class medicine'	great differences in treatment	high expenditure

The actual health care systems existing in different countries can be considered as combinations of (weakened) characteristics of these three ideal types. The universal governance problems in health care, therefore, have a threefold nature, based on the three governance problems of the ideal-type health care systems. The basic difference between the three systems consists in the goal the health care system should realise in addition to the shared goal of optimal medical care. In the so-called *guild-free-choice system* this additional goal is to maintain an optimal relationship of trust between physician and patient. In a *centrally planned national health service* this is a fair distribution of resources among the population, and for a system of *managed competition* it is the efficient allocation of (medical) resources. The different goals are attained in different institutional settings, which can be characterised by the criteria mentioned in the first column of Figure 2. The main differences between the three ideal types relate to the dominant level of decision making, macro, meso, or micro, and the way of coordination, namely by imperatives through hierarchical lines, by agreement between an autonomous medical profession and other groups, or by the market using prices and contracts. The *guild-free-choice system* forms the basis of health care systems in, for example, the Netherlands and Germany. A form of the centrally planned system is found in the United Kingdom, Canada, and Scandinavia, while in the United States a system of managed competition comes more and more into use.

We noted that each of the ideal-type health care systems has one or two additional goals. In the terminology of our analysis of medical practice we can say that each system emphasises one or two of the different kinds of constitutive rules. These rules define adequate performance or, in other words, the `standards of excellence' or quality standards for actions within medical practice that the system seeks to facilitate. So, in the context of a health care system these constitutive rules can be considered as quality aspects, and each of the theoretical systems emphasises a specific quality complex.

1) The *guild-free-choice* system puts emphasis on optimal quality of the `primary process' (especially the physician-patient relationship) in which two components can be distinguished: a physician's medical-professional competence (the scientific and technical aspect of quality), and a good relationship based on mutual trust between a physician and a patient (the social aspect of quality).

2) The system of *managed competition* gives priority to efficiency[12] (the economic aspect of quality of the medical care), though this does not imply that balanced attention is given to the four complementary dimensions that are distinguished in economic literature on efficiency[13].

3) The quality complex that is paramount in the *planned health care* system is the accessibility to and a fair distribution of health care [the judicial aspect of quality (distributive justice)].

Each of the systems also has its typical governance problems. (See the lowest row of Figure 2). In the centrally planned health care system quality can suffer due to the central decision making and the use of bureaucratic controls. In practice things can get even worse if the planned capacity is kept below the level of need. In this case a private upper class health care sector may emerge, which is in strong contradiction with the goal of a fair distribution of health care facilities. In the guild-free-choice system costs may increase in an uncontrollable way, so that health care becomes too expensive. Futhermore, if in practice the medical profession is less homogeneous than it theoretically should be, great treatment differences may occur. Finally, managed competition has no basis for solidarity nor for a trust relationship between physician and patient. Furthermore, the system is susceptible to upward pressure on expenditures, as the experience in the United States has shown.

The weak points of one particular system, however, often form a favourite quality aspect in one of the other systems. From the principle of the *simultaneous realization of norms*, this implies that in the `ideal' health care system justice is done to the three quality complexes at the same time. Adjustments of existing health care systems should, therefore, be evaluated on the consequences for these three quality complexes. This applies to both small adjustments and fundamental system revisions. System adjustments relate to minor changes which only introduce some elements of the other ideal-type systems into an existing system based on one of the ideal-type systems, while maintaining the essential characteristics of the actual system. These minor changes, which can have a major impact, are realized by so-called governance instruments. Fundamental revisions of health care systems in our terminology relate to a shift of all system characteristics from one ideal type to the neighbourhood of another ideal type. Thus, the introduction of managed competition can be seen as a fundamental system revision.

Governance instruments

Figure 3 shows a typology of governance instruments. Four alternative ways of governance and three levels of functioning are distinguished. These three levels of functioning are complementary, from macro level downwards. This means that in a centrally planned health care system governance instruments are used at all levels, mainly the instrument of imperatives. In a guild-free-choice system the meso and micro level instruments are complementary. A minor change in such a system can be the introduction of a macro budget for health care, as long as the profession maintains a dominant influence on the allocation of this budget. Similarly, the system undergoes a minor change if *ceteris paribus* a new remuneration system is introduced, without altering the physician's free choice of prescription, referral and treatment. According to the principle of the simultaneous realisation of norms, one may conclude that the most adequate health care system consists of a finest blend of governance instruments at all three levels of functioning. The choice of specific governance instruments, however, is of major importance, because these instruments not only (may) correct some problems of a particular system, but also affect medical practice. In order to establish whether those effects are ethically justifiable, an evaluation of a particular instrument is desirable before it is implemented. Our analysis of medical practice provides us with criteria to evaluate the various instruments.

Fig. 3. A typology of governance instruments according to functioning and level of functioning

	MACRO LEVEL health care sector	MESO LEVEL sectors and specialisms within the health care sector	MICRO LEVEL physicians and local institutions within sectors and specialisms
CLOSING AN 'OPEN END'	Macro budget for health care	Budgetting: - sectors - specialisms	Budgetting: - physicians - institutions
GOVERNANCE BY IMPERATIVES	Planning of total capacity of health care provisions Stratification of facilities (e.g. care vs cure)	Planning of: - sectors - specialisms	Planning of: - physicians - institutions 'managed care'
GOVERNANCE BY INCENTIVES		Remuneration system -sectors -specialisms	Remuneration system: - physicians - institutions Protocols peer review
PROFESSIONAL SELF-REGULATI-ON		Consultation be-tween and within specia-lisms Professional agreements between and within specialisms	

Evaluation

The general principle formulated in section 3 provides the first criterion for such an evaluation:
1) does the governance instrument respect the specific responsibilities and the related freedom of action of the actors involved, especially of the physician?

The survey of the three ideal-type health care systems indicates the quality complexes that are emphasized in each system, as well as the governance problems that each has. These strong and weak points of each of the systems provide a clue for the formulation of criteria for the ethical evaluation of governance instruments. This leads to the following criteria:

2) *what is the effect of the governance instrument on the medical scientific and technical quality of the medical care and on the confidence relationship between physician and patient?*
3) *what is the effect of the governance instrument on the efficiency and the overall costs for health care?*
4) *what is the effect of the governance instrument on the accessibility and distribution of care?*

With the aid of these criteria we evaluated one fundamental system revision, namely the introduction of managed competition, and three governance instruments: protocolling, budgetting and remuneration system. In order to be able to apply the criteria we have surveyed from literature the effects of these instruments on the performance and the freedom of treatment of the physicians. A description of our findings and of the evaluation of the instruments is beyond the scope of this paper that only deals with the main line of this project. Therefore, we will only present a few general conclusions.

1. The introduction of managed competition on prices in specialised curative health care is undesirable. The desirable professional freedom of treatment is threatened, the information gap between physician and patient hinders the patient in his free choice of the services that are offered and undermines the relationship trust with the physician, and the transaction costs of the necessary regulations will be very high.

2. Physicians have a freedom of treatment in view of their mission to primarily serve the interests of the patient. This freedom entails an important principle. Therefore professional self-regulation deserves to be respected and encouraged e.g. in the form of peer review, protocols, consensus meetings, etc. So we opt for the guild-free-choice system as basis for a health care system. However, this professional freedom and self-regulation requires that the profession is open to control and criticism and monitors the quality of professional care. The professional character of medical practice also implies that medical treatment requires a medical indication. This is an important principle in the context of cost containment in health care.

3. Budgetting is an effective and, as long as the budgets are based on medical needs, ethically justifiable governance instrument. To avoid an undesirable limitation of professional freedom, it seems advisable to give the medical profession a substantial responsibility for the expenditure of the budgets.

4. The fee-for-service remuneration system for medical specialists has several disadvantages. We would favour a kind of case payment - in as far as the problem of the pricing of the `cases' (patients with a certain diagnosis) can be solved satisfactorily - that for some specialisms could be combined with a basic income from a fixed salary.

We finish with a final conclusion of a different character from the previous conclusions.

Medicalisation and the growth of health care expenses are related partially to a predominant ethos in our society, namely an overestimation of health (often interpreted as a sense of well-being) and health care, and to a loss of possibilities to (meaningfully) integrate illness, handicaps and suffering into our concept of life. If modern society does not succeed in diminishing the expectations for a happy and healthy life from health care, health care expenditure will tend to increase excessively and an enforced cost containment will cause serious social problems.

Henk Jochemsen, Jan Hoogland, Johan Polder, Lindeboom Institute, Ede, Netherlands.

NOTES

[1] This term is derived from Williamson O.E., (1989), *The economic institutions of capitalism,* New York, who introduced the term governance structures. A governance instrument is a measure at the meso- or macro-level taken by the government or other actors in health care to bring about changes in the functioning of the health care system.

[2] In this paper the term `freedom of action' is used, which refers to an element of the broader term professional autonomy; with `freedom of action' we indicate the freedom of the physician in clinical situations to do that which from a medical point of view, is considered the best for the patient.

[3] MacIntyre A. (1981), *After virtue. A study in moral theory.* London: Duckworth, p.175.

[4] For the difference between goal and finality see: Dooyeweerd H. (1953- 1958), *A new critique of theoretical thought.* Amsterdam/Philadelphia: The Presbyterian and Reformed Publishing Company, vol. III p.570, 571.

[5] Strijbos S. (ed. 1992), *De medische ethiek in de branding - Een keuze uit het werk van Gerrit Arie Lindeboom*, Amsterdam ,p. 121, 122.

[6] See lemma `profession' in Webster's third new international dictionary.

[7] See for example Unschuld, PU. Professionalisierung und ihre Folgen. In: Schipperges H, Seidler E, Unschuld PU. (eds). Krankheit, Heilkunst, Heilung. Freiburg/München 1978: p.519,520. Unschuld concludes that the central meaning of the concept of profession is the pursuit of the selfless rendering of a public service and the independence in practicing the work typical for the particular profession.

[8] This ethical commitment of the medical profession is also formulated in medical ethical codes. For instance the Declaration of Geneva of the World Medical Association states: "The health of my patient will be my first consideration" and the Declaration of Helsinki says in its introduction: "It is the mission of the medical doctor to safeguard the health of the people".

[9] Cf. Puolimatka T. (1989), *Moral realism and justification.* Helsinki: Suomalainen Tiedeakatemia, p.143-154. Puolimatka defines benevolence as "a normative attitude which regards the well-being of others as intrinsically valuable", p.144.

[10] In this classification government-owned health care institutions belong to category 3.

[11] Figure 2 is based on literature on the characteristics of economic systems. See, for instance Gregory PR, Stuart RC, *Comparative Economic Systems.* Boston: Houghton Mifflin Company, 1989.

[12] In addition to efficiency arguments ideals of liberty are important, cf.:R.E. Moffit.(1994), *Personal freedom and responsibility: the ethical foundations of a market-based health care reform. The Journal of Medicine and Philosophy*, vol.19:471-81.

[13] These four dimensions are: technical efficiency (cost-effectiveness of the performance of a service), allocative efficiency (efficiency of allocation of service or provision to a person or to a need; "mainstream economics"), dynamic efficiency (renovation of materials and techniques by innovation leading to more effective or efficient treatments; "Schumpeter"), transaction efficiency (efficiency pertaining to the expenses of implementation of the health care system as organisation form; "Willamson")

PART 2.2

BIOETHICAL JUDGMENT: EPISTEMOLOGICAL STATUTE AND INSTITUTIONAL CONTEXT

ROBERTO DELL'ORO

INTERPRETING CLINICAL JUDGMENT: EPISTEMOLOGICAL NOTES ON THE PRAXIS OF MEDICINE

INTRODUCTION

Clinical judgment represents a topic of such complexity and importance that to pose the epistemological question in a direct fashion seems an act of either intellectual *hubris* or methodological *naivetè*. This is the case for at least two reasons. There is, first of all, the problem of defining the appropriate field for the topic. The vast body of literature dealing with issues of "philosophy and medicine" is still searching for methodological consistency and for the coherence of a systematic frame[i]. Sharing in the current post-modern cultural mood[ii], philosophers of medicine have become increasingly weary of comprehensive systems[iii]. Indeed, the "deconstruction" goes so far as to cast doubts on the plausibility of philosophy of medicine itself[iv]. According to Henk Ten Have the emphasis on bioethical issues has practically reduced the threefold spectrum of philosophical traditions in medicine -- the epistemological, the anthropological, and the ethical -- to the last one. As a result, "it seems that philosophy of medicine has come to an end, or that it has been transformed into bioethics."[v]

With that I mind I come to the second, no less powerful difficulty in delimiting the topic, namely, the plurality of scientific approaches of formal analysis to clinical judgment, and the problem of assessing the claim to objectivity exhibited by each one of them[vi].

The differences among competing theories -- from those based on syllogistic procedures to various probability theories, judicial algorithms, and Bayesian analysis -- run so deep as to imply a particular understanding of the meaning of medicine as a scientific enterprise and an implicit or explicit theory of human knowledge.
I submit that an epistemology of clinical judgment should not be reduced within the limits of a purely *formal* enterprise. Sceptical of such a reductive epistemology, I will fame the problem of an epistemology of clinical judgment within the broader context of the *praxis* of medicine, ultimately defined by the reality of the clinical encounter[vii].

My notes unfold along the line provided by three fundamental reflections. First, I lay out the essential features of clinical judgment as they appear through broad

C. Viafora (Ed.), Clinical Bioethics. A Search for the Foundations, 155-167.

phenomenological analysis. Secondly, I mention the interpretive nature of clinical judgment by focusing on the experience of illness and the physician-patient relationship. Finally, I address clinical decision making with a hermeneutic model of application.

PHENOMENOLOGY OF CLINICAL JUDGMENT

The thrust of my analysis is descriptive or phenomenological *latu senso*, in that it takes description of experience as the first step in any philosophical approach. The task at hand consists in describing the phenomenon of medical judgment, so as to grasp the meaning of the experience that leads the physician, through a process of argumentation and reasoning, to a particular therapeutic action on behalf of the patient[viii]. The presumption underlying this approach is that phenomena can speak to us if only we allow them to fully appear for what they are[ix]. Thus, describing is more than an empirical transcript, a pure recording of reality. Indeed, phenomenologists underline the *a priori* nature of "phenomenological experience," so that everything and anything that is given rests on experience. As for the *empirical* nature of this experience, consider the following observations of Max Scheler:

> "He who wishes to call this empiricism may do so. The philosophy which has phenomenology as its foundation *is* empiricism in that sense. It is based on facts, and facts alone, not on construction of an arbitrary "understanding." All judgments must conform to *facts*, and methods are *purposeful* only insofar as they lead to prepositions conforming to facts.[x]

The goal of phenomenological description is not to provide a copy of things, but to uncover those fundamental structures that make things what they are. The act of phenomenological description is, in the words of Dietrich von Hildebrand, a *prise de conscience,* an intellectual insight that transcends the realm of empirically verifiable factuality[xi].

Clinical judgment as action

I take the essential phenomenological feature of clinical judgment to be its dynamic, process-like nature. Whatever formal method is chosen to analyse the logic that leads to a particular medical judgment, one can relate the particularity of a single operation to a series of other operations. Indeed, clinical judgment must be considered within a larger contextual frame, spanning from the collection of medical data, to diagnostic and prognosis assessments, to the selection of a specific treatment; it is a continuous action distended along time. Of course, it is possible to isolate various moments in the process, and to analyse them in their own individual meaning. Yet, those individual moments constitute segments of a broader context that only as a whole can account for the reality of the clinical action. Within this broader context, each moment relates to the next; moreover, each moment necessarily moves in a *crescendo* that culminates in the medical decision. The latter represents the focal point, the situation of maximum ontic density toward which the other points of the clinical action converge. The broader phenomenological characteristic of the clinical phenomenon considered as a process is its intrinsic *teleology*: all the elements of medical judgment acquire their meaning in light of the final point, or *telos,* toward which they move.

Metaphysical difficulties historically raised against teleology impose a more nuanced reflection. In particular, the teleology underlying clinical action cannot be described as an *ideal* trajectory of unrelated points whose connection is being projected onto the action itself by an act of "synthetic apperception." Such a connection would not define the action in its ontological consistency, but only a function of the observer's mind. In Kantian terms, it would be only a transcendental condition on the side of the subject.

On the other hand, the presumption of phenomenological description is one of *ontic* realism. Something is seen as the objective correlate of an intentional relation, even if no conclusions are immediately drawn on the *ontological* nature of such intentional correlate. Yet, one might say that the process-like nature of clinical judgment *is* the reality of an action understood in relation to a situation of illness. The clinical decision represents the closure of a process in which all the elements are taken into account and synthesized around the final question, "What should be done for this particular patient?"

In his study *The Anatomy of Clinical Judgment,* Edmund Pellegrino takes this meaning of teleology much further, signifying not only the culmination of clinical judgment as a process, but the fulfilment of the medical encounter and, ultimately, of medicine *tout court*:

> "The end of the medical encounter, and the process of clinical judgment through which it is achieved... is restoration and healing – some corrective, remedial or preventive action is directed at what the doctor and the patient's wholeness, each in his/her own fashion. The end is not diagnosis, a scientific truth, testing an hypothesis or evaluating a treatment, though the knowledge derived therefrom enters into several states in making the decision to act"[xii].

Perhaps Pellegrino's use of teleological language commands a deeper articulation of different levels of discourse at work here. Suffice it to say that the reference to a decision that functions as the fulfilment of the clinical process is important: it clears the ground for understanding clinical rationality as *practical* rationality, one in which the explanatory model implied by the scientific understanding of disease is not suspended or superseded, but rather integrated as a dimension, however essential and important, of the larger *praxis* of medicine.

Scientific reasoning and the praxis of medicine

If the ontic specificity of clinical judgment is dynamic progression toward a *telos*, "a being which is becoming", in the words of Aristotle, then one must exercise a methodological *epochè*, a suspension of belief on any approach that could potentially compromise the specific nature of the phenomenon in question. The relation of intentionality itself requires such a suspension as it presupposes a structural homogeneity between subject and object, consciousness and experience, *noesis* and *noema.*

Phenomenological *epochè* has, first, a *negative* and critical meaning. Its function, rather than indicating the approach to take in order to grasp the essence of a particular realm of experience, is to clear the very ground of the subject's consciousness from any theoretical presuppositions that could potentially compromise the full affirmation of the phenomenon's *eidos*. In particular, the *epoche'* functions as a critical *caveat* against positivistic approaches to medicine and to scientific inquiry, approaches that ultimately define the general character of modern natural sciences.

In *The Enigma of Health*, Hans Georg Gadamer poignantly describes modern scientific knowledge as a capacity to produce effects: the mathematical-quantitative isolation of laws in the natural order provides human action with the identification of specific contexts of cause and effect; also, it empowers human action with new possibilities for intervention[xiii]. Gadamer's hermeneutic of the modern scientific enterprise echoes, in many ways, another critical account of the scientific idealization of experience. In his famous *Die Krisis der europäischen Wissenschaften*, Edmund Husserl provides a "genealogy" of experience that, insofar as it represents an experience of *Lebenswelt,* or life-world, precedes its being idealized by science[xiv]. Using a concept consciously formulated in contrast to a concept of the world that includes what can be made objective by science, Husserl calls "life–world" the world in which we are immersed in the natural attitude (*natürliche Einstellung)* that never becomes an object as such for us, for it represents the pre-given basis of all experience[xv].

The positivistic abstraction underlying the concept of technology in the modern scientific ideal acquires new and specific possibilities in the field of medicine and its healing procedures[xvi]. In relation to clinical judgment, such an idealization could be understood, in the first place, as a negation of the particular nature of the phenomenon, as a tendency to reduce the contextual *praxis* within which the clinical judgment takes place to the objectivity of theoretical knowledge; moreover, to interpret the healing process itself as a production of effects. Of course, the application of scientific reasoning to clinical judgment is not being questioned. In trying to determine what is wrong with the patient, in attempting to identify and explain the cause of symptoms, the physician does indeed deploy probabilistic laws and rules, theories and principles, of the biomedical sciences. Concepts of normal and abnormal, for an example, are statistically derived concepts based on scientifically validated norms of human biological functions. In the attempt to classify the patient's symptoms as manifestation of a particular disease entity, the physician relies upon the intrinsic possibilities of hypothetic-deductive reasoning.

Modes of scientific reasoning also define the therapeutic question. In trying to determine what can be done to remove or alleviate the cause of the patient's suffering, the physician appeals to prognostic knowledge about the course of diagnosed disease and about the efficacy and toxicity of relevant therapeutic possibilities.

Yet, *clinical* reasoning cannot be entirely equated to *scientific* reasoning. The goal of the former is not to relate different segments of scientific explanations to a unified theory. Rather, it is to bring together -- in a synthetic action that is theoretical and practical at the same time -- an understanding of illness with a specific medical decision on behalf of the patient. Unlike the patho-physiology of disease, the phenomenon of illness cannot be observed, analysed, and explained in itself. It must be understood as a part of the life-world of the subject in whom it manifests itself[xvii]. This is the reason why medicine represents a unity of theoretical and practical knowledge within the domain of the modern sciences, "a peculiar kind of practical science for which modern thought not longer possesses an adequate concept"[xviii].

Retrieving the subject: from action to the agent

A further fundamental step in a phenomenological inquiry of clinical judgment leads to the question of the *subject* (or the subjects) involved in the clinical action.

Originally influenced by positivistic ideals of science, analytic philosophy has defended the possibility of a philosophy of action in which the question of the subject is never uttered and, ultimately, completely ignored[xix]. The difficulty of this position is dramatically illustrated by the development of analytic philosophy itself, for which the "semantic of action" is pulled back to its inter-subjective condition of possibility, in a "pragmatic of action"[xx]. As Toulmin himself points out, this development can be seen in a wide range of analytic philosophers, from Wittgenstein, to Austin, and Searle[xxi].

Yet, positivistic prejudices and factual idealizations keep lurking behind the analytic skills of an "agent-less" theory of action. Within the general constraints of such a paradigm, it matters not who the subject of the action is: the action is treated simply as a sub-class of impersonal events.

In a series of articles collected in the volume *Actions and Events*[xxii], Donald Davidson presents a theory of action in which the distinctive teleological character of action is subordinated to a *causal* conception of explanation. Causal explanation serves, in its turn, to place actions within a general ontology in which events are understood as incidental occurrences, as irreducible entities placed on the same level of substances as fixed objects. This ontology of "impersonal events" ends up structuring the entire gravitational sphere of the theory of action, preventing an explicit, thematic treatment of the relation between action and agent.

In light of an impersonal ontology of events, clinical judgment could very well be interpreted as the result of a computer based operation that depends exclusively on the completeness and accuracy of the information being submitted[xxiii].

One can see that the increasing specialization of medicine as a discipline, the anonymity of hospital procedures, and the powerful influences of economical forces operating behind the health care industry contribute to slowly, but surely, concealing the reality of a personal agent in medicine. And yet, the question of the subject cannot be ignored in a phenomenological analysis of experience. Here, the symmetric polarity of subject and object must be interpreted -- by the necessity of an essential connection -- within the frame of an intentional relation. How does the question of the subject become relevant in the specific case of the action being at issue here, namely the clinical judgment?

Intentionality is a term that refers originally to the theory of knowledge[xxiv]. It underlines the fact that the consciousness of the knowing subject does not exist in-itself, prior to its relations to an object, but it is always object-oriented, i.e., consciousness *of* something. Correlatively, the object never exists as an object-in-itself, but always as a correlate of a consciousness. Along the same line, philosophers working in phenomenology have uncovered the intentional dimension of other operations in the subject. For an example, Max Scheler points to the intentional meaning of emotions and feelings, bringing forth their deeply personal and spiritual dimensions against deterministic and materialistic hermeneutics of instinct[xxv].

The application of the notion of intentionality to the realm of action, however, represents a kind of extension in the phenomenological theory. In fact, it provides a framework for the interpretation of clinical judgment within the context of a *personal* relation; better, in terms of a personal encounter.

The intentional dimension of action can be described retrospectively. It implies going up-stream against the intentional flux, in order to reach out for the agent, or the agents, involved in the action. In this light, an action always proceeds from somebody and is directed at somebody else. Looking at the action intentionally means, therefore, overcoming an objectivistic attitude that rests, *de facto*, upon the separation between subject and object.

An action is never just an impersonal state of affairs out-there-in-the-world. There is something ambiguous about the language of bioethics when reference is being made to the "puzzle" posed by clinical cases[xxvi]. Taken by itself, a case represents an abstraction, an objectification that extrapolates from the intentional context, or the life-world of meaning and experience within which it is always embedded. In the words of Gadamer:

> "the concept of life-world is the antithesis of all objectivism. It is an essentially historical concept...(it) means the whole in which we live as historical creatures... It is clear that the life-world is always at the same time a communal world that involves being with other people as well. It is a world of persons, and in the natural attitude, the validity of this personal world is always assumed"[xxvii].

For this reason, the context of clinical judgment is always a personal context. Which is to say that a case is always somebody's case: somebody's life, but also somebody's responsibility and conscience in dealing with the complexity of the situation are more than accidental variables in the circumstantial texture of the case. They are the very stuff of which the case is made. In the same fashion, a particular decision on behalf of the patient is not just a strategic solution, or a technical fix to the complexity of an anonymous incident. From a phenomenological point of view, that action represents, first of all, the *actualisation* of a subject, a person's practical involvement whose effect is more than a change in reality. Indeed, it represents a change in the subject's experience, a modification of the subject's *being-in-the-world.*

The change pertains primarily to the subject who is a *patient.* The modification brought about by the physician in restoring health cannot be adequately described simply as the production of a biological state of affairs; rather, it represents the re-composition of a natural equilibrium whose essential features extend to the life-world of the patient[xxviii]. Just as sickness represents a situation of *dis-ease,* a rupture and break in the position of the human individual within the totality of being, so restoration of health predisposes the ground for a new personal synthesis, the re-unification of a life-world previously shattered or compromised by the event of illness.

Yet, the change is no less radical for the *physician* than the patient. Precisely because the physician's action is more than a production of effects, it demands from him/her not just the application of technical skills; the physician's life-world is put in question as well. This is, in my opinion, the radical meaning of the notion of care and the very ground upon which an ethics of care rests[xxix]. Of course, the good physician is one who cares for the patient in a *moral* sense, in so far as he/she empathizes with the patient's situation and frames his/her action in relation to this situation. Yet, care pertains to the quality of the action only because it defines the ontological condition of an acting person. It is the reality of a personal actualisation continuously challenged and kept in motion by the intrinsic demands of a relationship what gives meaning to the clinical experience and to its different dimensions.

CLINICAL JUDGMENT AS INTERPRETATION

The phenomenology of clinical judgment ends with an attestation of subjectivity. Even if the epistemological importance of such attestation needs to be further determined, the reference to an inter-subjective polarity at the heart of clinical judgment seems to lead in the direction of interpretation. Indeed, the process of clinical judgment is defined by interpretation from the beginning to the end.

The patient's experience of illness

This can be seen, first of all, in relation to the way in which a patient understands illness and seeks the help of a physician. As any other human interaction, the clinical encounter rests on the freedom of the persons who enter in relation with one another. Ideally, a patient partakes of a particular healing relationship without being forced into it. This holds true independently of the concrete availability or choice of physicians.

Correspondingly, the physician accepts the reality of the patient on the basis of a personal commitment, a promise to help, itself predicated upon freedom. Clinical medicine is the inter-subjective exchange in which an individual in need of healing entrusts himself/herself to another individual who professes and promises to heal on the basis of acquired knowledge, skill, and experience[xxx].

There are, at the same time, deterministic dimensions to this relation that are rooted in the very reality of illness. There are very good reasons for referring to the *fact* of illness as the first, constitutive element of medicine, if illness is indeed something that *happens* to a person. Even when a particular disease can be traced back, etiologically, to a certain life-style, one never chooses the particular suffering that comes with the ailment. By definition suffering entails an element of passivity (*pati*), which is absolutely personal. In the words of Paul Ricoeur,

> *"La soufferance est, avec la jouissance, la retraite ultime de la singularitè"* [xxxi].

We may suffer of the same disease, yet we undergo the experience of suffering in different ways, radically left to our own individuality. Even as empirically reducible entity, the fact of illness exists only as interpreted fact, experienced and recounted by a particular patient. Consider how, from the moment in which medical data are collected, the subjective perception of facts takes central place. Clinicians comment often on patients' tendency to inadvertently shift their language from the pure enumeration of symptoms to a kind of self-inferred diagnosis. The phenomenon might be explained in a variety of ways; not ultimately, the psychological need to control and define what one experiences. Insofar as it represents a spontaneous tendency, however, it throws into relief our need to interpret illness, to "reduce" its brute facticity to meaning[xxxii].

The subjective dimension of illness, including the transcendental function of interpretation, leads to two additional observations. Illness might not be associated with demonstrable pathology. Likewise, pathology might be present even when the patient does not experience himself/herself as sick. These observations presuppose understanding the distinction between illness and disease: the former referring to the subjective experience, and the latter to the objective construct.

Even if the distinction between illness and disease is basic to the critique of a model in which medical objectivity is seen as a "flight from interpretation"[xxxiii], it is no less important to avoid separating the two notions as completely foreign to one another. Indeed, the separation would simply legitimise an approach to illness, as well as an understanding of medicine, totally unrelated to their subjective and, therefore, interpretative variables. Interpretation represents the proper epistemological mode for understanding illness because our very access to reality is structurally mediated by an act of interpretation. In this sense, human understanding is hermeneutic at root[xxxiv].

By looking at the patient, I have shown how the intentional dimension of clinical judgment shapes the subjective perception and definition of the experience of illness. The interpretative nature of this experience becomes evident in the language used by the patient, in the emotional mood underlying his/her narrative of symptoms and pain. It is important to stress that narratives of illness are never purely descriptive: insofar as they are embedded in the life-world of patients – bespeaking their beliefs, fears, uncertainties – they are already value-laden. In their narratives, patients are already trying to understand; in the process of understanding, they are also interpreting.

Physician's attitude: listening and caring

At the other side of the relation, the physician listens to the patient's story. There is something very profound about this attitude of listening that directly affects the clinical judgment. Indeed, hearing what the patient has to say conditions the clinician's ability to understand the phenomenon of illness[xxxv]. More than just a professional virtue, listening represents for the health care professional an ontological specification, or, in the words of Martin Heidegger, a fundamental existential possibility:

> "If we have not heard 'aright,' it is by no accident that we say we have not 'understood.' Hearing is constitutive for discourse. Listening to is Dasein's existential way of Being-open as Being-with for Others... Dasein hears, because it understands"[xxxvi].

For the patient, as for the physician, illness becomes the intentional correlate of an act of interpretation. For this reason, in the diagnostic process, the physician needs to rely on the patient's story. Disease is never just a thing-in-itself, ready to be grasped in a scientific act of detached objectification, predicated upon the "suspension" of the patient's narrative, experience, and history. On the contrary, the objectification of disease cannot be separated from the subjective account of illness. Indeed, it is *through* the interpretative nature of the patient's story that the physician will eventually explain the objective nature of illness. Just as disease is embedded in the experience of illness, so scientific judgments of its objective nature represent derivative modes of interpretation, grounded in the reciprocal understanding made possible by the clinical encounter. With a term borrowed from hermeneutics, one could re-express the kind of understanding that occurs between the physician and the patient in terms of a fusion of horizons (*Horizontsverschmälzung*). The horizon represents the pre-comprehension structuring —from different perspectives—the understanding of illness[xxxvii].

The category of fusion of horizons stands dialectically against the notion of the clinical gaze as an act of objectification[xxxviii]. Of course, the ultimate goal of the clinical encounter is to isolate the cause of illness, by objectifying the etiology of the disease, not to establish a personal relation. In this sense, the "fusion" just referred to is more a means toward an end, than an end in itself.

Yet, the scientific objective of medicine can be achieved only through the intrinsic possibilities and difficulties of dialogue. For this reason, the diagnostic process does not represent a moment apart from the contextual reasoning that makes the communication possible. Such reasoning is *practical*, for it takes place within the *praxis* of communication defining the physician-patient relationship, and it is functional to achieving the *telos* of that relation. The inter-subjective nature of the physician-patient relationship does not add a new element, a purely external one, to the properly "scientific" side of the clinical judgment, as if the latter could stand independently of the relation within which it takes shape. Rather, the opposite is true: the relation itself intrinsically structures and defines the scientific side of the process in its very meaning.

PHRONESIS AND APPLICATION IN CLINICAL REASONING

With the question "What should be done for this patient?" clinical reasoning comes to closure. In determining a particular course of action for a particular patient, the clinician brings the process of interpretation of illness, grown out of the clinical encounter, to fulfilment. The understanding of illness as subjective experience finally translates into therapeutic intervention. The interplay of theoretical and practical dimensions, of objectivity and subjectivity structuring clinical judgment and, with it, the process of clinical reasoning, provides the framework for understanding the meaning and the importance of the clinical decision. I have argued that the need to contextualize the patho-physiology of disease within the patient's life-world is based on the expectation that the physician will *act* upon the patient as subject, rather than intervening on the biological entity as an object-in-itself.

Unlike the pure scientist, the clinician's job ends with the practical application of knowledge. The clinician's interest in the disease is ultimately oriented toward making the *right* decision on behalf of the patient. The epistemological significance of this last statement depends entirely on our understanding of application[xxxix].

Two competing models – which I will call scientific and hermeneutic – are available. In the *scientific* model, theory and praxis describe two entirely separated levels of reality. Indeed, separation represents the condition whereby theory is established in its universality, gaining validity through progressive abstraction from any particular case; theory stands independently of praxis. Application, understood as a function of verification or falsification, makes possible the passage from the general to the particular, from the objective to the subjective. Yet, such a passage functions only insofar as particular cases fit the hypothetical model of general explanatory principles. From this perspective, the rightness of a clinical decision is formally defined by its conformity to pre-existing parameters. What the clinician requires is purely the skill of the craftsman who learns to apply scientific knowledge and discoveries with the purpose of restoring health.

The *hermeneutic* model, on the other hand, recognizes the structural interplay of theory and praxis at the heart of clinical reasoning. Here the theoretical and the practical are not separated levels of reality, but polar dimensions whose validity is reciprocally co-determined. Far from being a purely mechanical function, application represents a true *mediation* between the universal and the particular, the theoretical and the practical. In this light, the theoretical understanding of illness is not gained independently of its practical manifestations and its subjective interpretation, but rather through them. Indeed, the rightness of a particular clinical decision will represent a synthetic mediation of universal and particular, whereby the general laws of diagnostic procedures are understood and interpreted in relation to the specific situation of illness affecting a specific patient.

It is impossible to miss the analogy between the hermeneutic model of application and the Aristotelian notion of *phronesis*[xl]. In Book VI of the *Nicomachean Ethics*, Aristotle distinguishes *phronesis* from the "intellectual virtues" of *episteme* and *techne*. In the same way here the notion of application as mediation is gained dialectically, namely, by playing out its hermeneutic meaning in contrast with the scientific model of application. The analogy reveals a form of reasoning appropriate to the specific praxis of medicine[xli].

In particular, Aristotle's *phronesis* seems to account for two important features of clinical judgment that were thrown into relief by the previous phenomenological description. The first concerns the *practical* nature of clinical judgment, the mediation it establishes by means of deliberation and choice between the universality of scientifically validated laws concerning the patho-physiology of disease and the particularity of concrete, i.e., personalized phenomena of illness. The second feature is the *intentional* dimension of clinical judgment, its necessary correlation to a subject who actualises himself/herself in the very process of judging. *Phronesis* is not to be confused with *episteme,* with an "objective knowledge" that is detached from one's own being and becoming. Just as *phronesis* (prudence) determines what the *phronimos* (the prudent person) becomes, so the application of clinical judgment progressively shapes the clinician into a moral agent. Conversely, it is the *phronimos* who – by relying upon experience and practice – is most likely to make the right decision and to fully account for the particular features of the case at hand.

CONCLUSION

These last observations seem to indicate a direction rather than a conclusion. My notes would probably have to abandon, at this point, their general epistemological concern and become more clearly ethical in tone. But this seems required by *die Sache selbst,* the very nature of medicine as praxis of healing. In the end, to interpret clinical judgment means to recognize the practical nature of medicine. In my analysis, I have shown that such recognition depends upon the intrinsic teleology of clinical judgment, the interpretive character of the physician-patient relationship, and the "phronetic" application of general principles of diagnostic intervention to particular instances of illness.

Roberto Dell'Oro, Loyola Marymount University, Los Angeles, CA.

ENDNOTES

[i] I see two different issues here. The first concerns the need and the plausibility of an overarching frame in a foundationalist fashion. The second question pertains to the epistemological status of philosophy of medicine as a separate discipline. See Edmund Pellegrino, "Philosophy of Medicine: Towards a Definition, *The Journal of Medicine and Philosophy* 11 (1986): 9-16, and more recently, idem, "What the Philosophy of Medicine Is", *Theoretical Medicine and Bioethics* 19 (1988): 315-336.

[ii] For an interpretation of postmodernism as mood, or *Stimmung* -- the term is, of course, Heideggerian in tone -- see Richard J. Berstein, *The New Constellation: The Ethical-Political Horizon of Modernity/Postmodernity* (Cambridge, Mass.: The MIT Press, 1992).

[iii] According to Stephen Toulmin: "This movement shared, at least, the conviction that all earlier quests for a comprehensive system of knowledge, based on permanent, universal systems of overarching principles, were misguided from the start, and are by now discredited. Claims to philosophical universality and permanence can be ignored: their only interest lays in the ways that they could serve as a "cover" for the collective interests of the nations, social groups, or genders with which their authors -- novelists, philosophers, or play writers -- were affiliated," "The Primacy of Practice: Medicine and Postmodernism" in Ronald A. Carson and Chester R. Burns, ed., *Philosophy of Medicine and Bioethics: A Twenty Years Retrospective and Critical Appraisal* (Dordrecht: Kluwer Academic Publishers, 1977) , 41-42.

[iv] See Arthur L. Caplan, "Does Philosophy of Medicine Exist", in *Theoretical Medicine* 13 (1992): 66- 77.

[v] Henk Ten Have, " From Synthesis and System to Morals and Procedure: The Development of Philosophy of Medicine" in R. A. Carson and C. R. Burns, ed., *Philosophy of Medicine and Bioethics,* op. cit., 107.

[vi] Alvan R. Feinstein, *Clinical Judgment* (Baltimore, MD: Williams Wilkins, 1967). Also H. Tristram Engelhardt, Stuart F. Spicker, and Bernard Towers, *Clinical Judgment: A Critical Appraisal* (Dordrecht: D. Reidel Publishing Company, 1979).

[vii] On this very point Marx Wartofsky, "What Can the Epistemology Learn from the Endocrinologists? Or is the Philosophy of Medicine Based On a Mistake?" in Carson and Burns, ed., *Philosophy of Medicine and Bioethics,* op. cit., 55.

[viii] I am aware of the fact that my analysis presupposes a particular type of clinical encounter as paradigmatic, one in which an ill person seeks diagnosis and treatment through a single health care professional. The physician- patient relationship may, of course, assume other forms that are as plausible and valid.

[ix] This may sound like a contradiction considering the claim to *Voraussetzungslosigkeit* or "suspension of presuppositions" invoked by phenomenologists. Indeed, the decision to suspend judgment already involves a fundamental ontological judgment on the formal intelligibility of reality.

[x] Max Scheler, *Formalism in Ethics and Non-Formal Ethics of Values* (Evanston, Northwestern University Press, 1973), 51-52.

[xi] Dietrich von Hildebrandt, *What is Philosophy?* (London: Routledge, 1991). The expression itself, however, is of Jacques Maritain. See the observations of Josef Seifert, *Erkenntnis objektiver Wahrheit: Die Transzendenz des Menschen in der Erkenntnis,* 2nd edition (Salzburg: Pustet Verlag, 1976), 28.

[xii] Pellegrino, "The Anatomy of Clinical Judgments. Some Notes on Right Reason and Right Action", in Engelhardt, Spicker, and Towers, *Clinical judgment,* op. cit., 172. Also Italian philosopher of medicine Paolo Cattorini, "Sulla Natura della Bioetica: una nota epistemologica sull'applicazione dell'etica alla scienza e alla clinica", in Elio Sgreccia, Vincenza Mele, and Gonzalo Mirando, ed., *Le Radici della Bioetica* (Milano: Vita e Pensiero, 1996), vol. 1, 77-83.

[xiii] H. G. Gadamer, *The Enigma of Health: The Art of Healing in a Scientific Age* (Stanford, CA: Stanford University Press, 1996), 35.

[xiv] See in particular Edmund Husserl, *The Crisis of European Sciences and Transcendental Phenomenology: An Introduction to Phenomenological Philosophy,* (Evanston: Northwestern University Press, 1970). For a general overview, Paul Janssen, *Edmund Husserl: Einführung in seine Phänomenologie* (Freiburg: Verlag Karl Alber, 1976).

[xv] For a general hermeneutics of the notion of *Lebenswelt* in Husserl see Paul Janssen, *Edmund Husserl,* op. cit., 135-145, Maurice Natanson, "The Lebenswelt", in Erwin Straus, ed., *Phenomenology: Pure and Applied,* (Pittsburgh: Duquesne University Press, 1964), 75- 104. Also Hans Georg Gadamer, *Truth and Method,*, 2nd edition (New York, The Continuum Publishing Company, 1994), 242-254; idem, "The science of the Life- world, " in Anna-Teresa Tymieniecka, ed., *Analecta Husserliana,* vol. II (Dordrecht: D. Reidel Publishing Company, 1972), 173- 185.

[xvi] For an historical account, see S. J. Reiser, *Medicine and the Reign of Technology* (Cambridge: Cambridge University Press, 1978). From a more theoretical perspective, Salvatore Natoli, "La costituzione dello sguardo medico: dal gesto terapeutico alla scientificità della medicina", in *Nuovi Saggi di Medicina e Scienze Umane* (Milano: Istituto Scientifico San Raffaele, 1985), 31- 70.

[xvii] See Edmund D. Pellegrino, "The Lived World of Doctor and Patient: A Phenomenological Perspective on Medical Ethics", lecture at Yale University, April 11, 1996, *Bioethics and Public Policy Symposium.*

[xviii] H. G. Gadamer, *The Enigma of Health,* op. cit., 39.

[xix] See for instance, A. I. Melden, *Free Action* (London: Routledge and Kegan Paul, 1961) and Stuart T. Hampshire, *Thought and Action* (Notre Dame, IN: University of Notre Dame, 1983). Also Arthur Danto, *Analytic Philosophy of Action* (Cambridge: Cambridge University Press, 1973).

[xx] On this question see the observations of Paul Ricoeur, *Oneself as Another,* translated by Kathleen Blamey (Chicago: The University Of Chicago Press, 1992), 27- 55.

[xxi] "Wittgenstein's focus in the *Tractatus Logico-Philosophicus* on "propositions" (*Sätze*) shifted, in the *Philosophical Investigations,* to *Sprachspielen* ("language games") and ultimately to *Lebensformen* ("forms of life") as the occasion of language games. What this shift made clear is that words, sentences, and other lexical items are not connected to their occasions of use by formal, logically necessary relations... J.L. Austin reminded us how many utterances operate more as *performances* than as representations of facts *(Bilder der Tatsachen*). John Searle similarly argued that they gain a meaning not in the way mathematical formulae do—by formal definition—but as other human acts do, from their roles in larger constellations of *human action.* Linguistic utterances differ from human gestures and songs, that is to say, only in being "speech" acts, rather than acts of other kinds", Stephen Toulmin, "The Primacy of Practice", op. cit., 45.

[xxii] Donald Davidson, *Essays on Actions and Events* (Oxford: Clarendon Press, 1980).

[xxiii] A computer-assisted diagnosis is part of a larger trend in medicine leading to what Wartofsky calls the "technologization of the medical subject". See "What Can the Epistemologists Learn", op. cit., 61. Also Luigi Stella and Antonella Crescenti, "Principi di Informatica clinica", In *Nuovi saggi di Medicina e Scienze Umane* (Milano: Istituto Scientifico H San Raffaele, 1985), 293- 360 and Elliot Sober, " The Art and Science of Clinical judgment: An Informal Approach", in *Clinical Judgment,* op. cit., 29- 44.

[xxiv] In the fifth of his *Logical Investigations,* Husserl elaborates the nature of intentionality. Even if the concept of intentionality is given a definite elaboration in *Ideas,* here already consciousness is understood not as an "object", but as an essential coordination. This constitutes an important starting point in overcoming objectivism. See the observations of Gadamer in *Truth and Method,* 244.

[xxv] Max Scheler, *The Nature of Sympathy,* translated by Peter Heath (London: Routledge and Kegan Paul, 1954). Also Paul Ricoeur, *Freedom and Nature: The Voluntary and the Involuntary,* translated by Erazim V. Kohàk (Evanston: Northwestern University Press, 1996).

[xxvi] For the notion of "puzzle" see Thomas Kuhn, *The Structure of Scientific Revolution,* 2nd edition (Chicago, University of Chicago Press, 1970), 36.

[xxvii] H. G. Gadamer , *Truth and Method,* op. cit., 27.

[xxviii] For Gadamer the understanding of human health in terms of the natural condition of equilibrium (*Gleichgewicht*) implies a more fundamental understanding of nature as equilibrium: "If we presuppose this idea of nature, then medical intervention must be understood as an attempt to restore an equilibrium that has been disturbed", Hans G. Gadamer, *The Enigma of Health,* op. cit., 36.

[xxix] I am aware of my not so implicit reliance upon Heidegger's hermeneutics of *Sorge* as ontological-existential category, i.e., as defining the being of Dasein. Such reliance does not imply, however, a negation of the specific ethical meaning of care. For an overview of the discussion on the ethics of care see the accurate reconstruction of Warren T, Reich in the *Encyclopedia of Bioethics,* 2nd edition (New York: Simon and Schuster MacMillan, 1995) vol. 1, 319-343.

[xxx] E. D. Pellegrino, " The Healing Relationship: the Architectonics of Clinical Medicine", in Earl Shelp, ed., *The Clinical Encounter: the Moral Fabric of the Physician-Patient Relationship* (Dordrecht: D. Reidel Publishing Company, 1983), 153-172.

[xxxi] Paul Ricoeur, «Le trois niveaux du jugement mèdical » *Esprit,* 12 (1996) : 21-33, at 22.

[xxxii] Indeed, the patient's narrative has not only a diagnostic, but even a therapeutic significance. According to Drew Leder the very ability to bring the disease to the level of language "counteracts two primary features of illness that give rise to suffering: senselessness and isolation... The fact of translating disease into a language begins to overcome this twofold alienation. What was a private pain is now made public, what was senseless and random is woven into a meaningful tale. The narrative context itself can

have healing force, " Drew Leder, "Clinical Interpretation: The Hermeneutics of Medicine," *Theoretical Medicine* 11 (1990): 9-24, at 13.

[xxxiii] See Drew Leder, "Clinical Interpretation", op. cit.

[xxxiv] Martin Heidegger, *Being and Time,* trans. by John Macquarrie and Edward Robinson (New York: Harper and Row, 1962), 182-203. On the importance of Heidegger's and Gadamer's philosophy in understanding this claim see Richard J. Berstein, "From Hermeneutics to Praxis", in *Philosophical Profiles: Essays in a Pragmatic Mode"* (Philadelphia: University of Pennsylvania Press, 1986), 94-114: "Implicit in Heidegger and explicit in Gadamer are two central claims: the ontological primacy of hermeneutics and its universality. We are thrown into the world as beings who understand; and understanding itself is not one type of activity of a subject, but may properly be said to underlie all activities," at 96.

[xxxv] See Richard Baron, "I Can't Hear While I Am Listening," *Annals of Internal Medicine* 103 (1985): 606-611.

[xxxvi] Heidegger, *Being and Time,* 206.

[xxxvii] In the process of understanding, a real fusion of horizons occurs -- which means that as the historical horizon is projected, it is simultaneously superseded. To bring about this fusion in a regulated way is the task of... historically effected consciousness." H. G. Gadamer, *Truth and Method,* 307.

[xxxviii] M. Foucault, *The Birth of the Clinic: An Archeology of Medical Perception* (New York: Vintage Book, 1975), 107-123.

[xxxix] My observations are grounded in Gadamer's analysis of application in *Truth and Method,* especially 307-311.

[xl] The hermeneutic relevance of Aristotle is, of course, recognized by Gadamer himself in *Truth and Method,* 312- 324. For a commentary, see Richard J. Bernstein "From Hermeneutics to Praxis", op. cit. and Paul Schuchman, "Aristotle's Phronesis and Gadamer's Hermeneutics," *Philosophy Today* 23 (1979): 41-50.

[xli] For an understanding of *phronesis* as a paradigm of clinical rationality see the study of Daniel Davis, "Phronesis, Reasoning, and Pellegrino's Philosophy of Medicine," *Theoretical Medicine* 18 (1997): 173-19.

PIERRE BOITTE

FOR AN ETHICAL FUNCTION IN HOSPITALS[1]

Experiences, opinions, remarks, reflections and questions about the existence of ethical committees within hospitals[2] show the emergence of an ethical function in hospitals. The presence of these ethical resources is recognised and necessary. But at the same time it is important to admit that the existence of this function must not just be confined to the functioning of ethics committees. To have an ethical function does not mean that we must have committees in hospitals (they are not always legally compulsory as they have been in Belgium since 1994 for example) as this ethical function could be functioning without the presence of committees.

So in this text a further reflection will help us to understand why an ethical questioning must exist and work (1), insisting on the critical role of caring experience, out of which an ethical reflection can come (2). The institutional functioning and the violence there is in it, make it difficult to set such a function (3) but this difficulty must be fought by accepting the positive role of the powerful relations set in the resistance to that violence (4). So we can say that the ethical function has a critical role, which enables to be creative and to set an autonomous strategy to ethics (5).

1- The aim of an ethical function in hospitals
2- The critical function of the experiences of the professional bodies
3- The institutional functioning
4- Towards relations of power without any violence
5- The ethical function as critical role

THE AIM OF AN ETHICAL FUNCTION IN HOSPITALS

Nowadays in hospitals there is a generalised ethical reflection which does not only come from the great evolution of the medical data nor from the clinical reality which integrates an ethical dimension. There is something more.

There are different reasons why the ethical reflection spreads out. On one hand the medical practices are more and more the subject for critical reflections (from sociological, economical anthropological approaches, institutional analysis, health policies). On the other hand critical reflections come from the practitioners themselves. Actually significant mutations in the occidental biomedical pattern show some increasing misunderstandings first between the aims medicine and institutions have and then between the finalities patients and professional bodies who do not indeed always understand the institutional functioning as they have less and less power on it.

In a way we can say that the development of the practice stays quite tight to the relevance of the critic as the internal critic of the carers is proving too marginal to be able to really change the systems of caring. But considering the inside and outside

C. Viafora (Ed.), Clinical Bioethics. A Search for the Foundations, pp. 169-180.

contestations it can also seem necessary to have a critical reflection in the spreading out of the medical practice, without however restricting it to cares but considering it in its social and institutional aspects, and this is our choice.

Such a critical reflection, as outside to the practice, is imperative if we want to avoid a normative regulation of the practices by logic of self-legitimation ("if a medical help to procreation exists, it must be regulated" as it is often said for example) This normative regulation would happen to the detriment of an analysis of the complexity of the concerned reality and at the risk of not taking into account the meaning of this complexity. *To defend the role of an ethical function in hospitals means in this perspective to agree to take into account this critical reflection in the biomedical practices, considering at the same time the outside part of the critical reflection, the proximity of actual caring and research practices and the institutionalisation of those two activities.* The bet is that the necessary critic of the biomedical rationality will lead not only to a clear analysis of the ethical issues of improvement due to that rationality but also to a true debate about the representations inherent to these improvements. We think that the people who are involved in the spreading of biomedicine and who are on the way to be more conscious of the ethical dimensions of their practice can have a critical reflection thanks to this debate.

The ethical function in hospitals consists in introducing the complexity of the reflection within the contemporary medicine which is tempted to develop a scientific and technical reason which will be reductionist and objectifying. The risk is that in its normative pretension this medicine would set up as determining what is right to be done on bodies and minds. The objectifying reflection is not bad in itself but what is to be fought is the only univocal interpretation of the causality which would prevent another possible meaning (of the illness for example) different from the one given by medicine (as we know that the ethical questioning is possible only if the meaning is left open and its interpretation left plural).

The critical reflection does not consist in formulating regulations in order to try containing or bringing limits to practices (which very often do not take care of those normative processes) but more in promoting the commitment of the persons who are fully acting in biomedical innovations, and specially those who practice medicine and care people. That is why in relation to these practices we have at the same time a close and far position. That position has been developed for several years in the Centre d'Ethique médicale at the Catholic University in Lille.

To succeed, such an approach must respect three requirements:

1- first, to help our reflection we must take into account the ethical capacity of professional bodies, often seen in the way they protest or say they are perplexed when caring
2- then, from this ethical capacity they have, we must start a dialogue with a theoretical ethical reflection
3- finally, we must put into words the questioning of the carers so that they do realise that they are social and political subjects being involved in the society debate[3].

THE CRITICAL FUNCTION OF THE EXPERIENCE OF PROFESSIONAL BODIES

The most common way to consider the ethical problem in the biomedical practice consists in working out ethical helping method to the decision taking. Facing a difficult situation full of perplexity, a choice which will be discussed in the team has to be taken so that a suitable action can be done and evaluated. But a problem appears when among the carers and in spite of those methods some of them are not satisfied because they have the feeling that they have not coped with the situation. Other questions come out then: (i) What has been done, has it been for the good of the patient? (ii) Have we been able to help him or her morally? (iii) What is the true meaning of our work?

Beyond the help to the decision taking, it is often interesting to propose an ethical and philosophical reflection from what the carers have felt from their experience. This will help them assume their practice in a critical way without forgetting their perplexity. From the experience of the professional bodies, from what is understood of it, we must then make them understand the philosophical and ethical aims of their practice.

In fact listening to the medical bodies talking about their practice shows that their experience is often felt as uncertain and perplexed, due to the difficulties coming from different possibilities offered by biomedicine today. Those difficulties are fed by the importance of the scientific and technical context, by the emotion felt in front of distressing situations, by the complexity of argumentation for the moral discernment, by the uncertainty as for the prediction of individual or collective consequences of choices. The carers cannot always assume those responsibilities which seem excessive. This notion of excess seems to characterise the caring experience. The difficulty in using ethical creativity seems an essential part of this experience. Many professional bodies have the impression that they have to assume alone the difficulty in taking a punctual and singular decision in a context of fragile and moral points of reference which makes the ethical discernment more important and more difficult.

Then the ethical reflection as an essential dimension of the experience of a subject can be done, if we accept that the experience is unique to the subject and it enables him or her to understand himself or herself from and through what has happened to him or her. The involvement in a caring practice leads the carers to a process of understanding themselves through their will to understand singular events in their practice. The ethical reflection helps them stand back on their own convictions and on the social function as carers; it also helps them have a better understanding of themselves as it lets their ethical subjectivity come out and as it lets their capacity give reason to their act.

Two parts of this experience which cannot be separated have to be considered. On one hand the involvement of the carer puts him or her in face of the experience of illness of the other, and to the potentialities and uncertainties of medicine. This facing gives him or her an excess of responsibility. On the other hand all those confrontations are part of the own experience of a carer. Facing the suffering of the other, the human finiteness and fragility which weaken his or her own fundamental convictions, the carer has to explicit his or her conception of the human being. Reflecting about his/her experience enables him/her to strengthen his/her critical ethical capacity. Moreover, this reflection takes into account this experience as a way to work on the ethical capacity and the possibility of assessing his/her professional identity.

In such a perspective, the caring experience is assigned a critical position. We are not going to have an ethical reflection because we set up groups of reflection or committees of regulation in which the part of experts is essential and in which the carers are rarely considered as important actors, though they are the ones interested by this reflection. A true critical position stands on the appeal of the caring experience. This appeal allows the involvement of the caring actors in their necessary critical task about the caring practice.

The ethical reflection, the development of an ethical function in hospitals must sustain the carers in what is their reality in the professional practice. It means accepting the complexity of this reality without willing solving it but at the same time knowing that it will not alienate their ethical capacity. So they can always question the logic of medical caring and institutional practices.

The ethical questioning comes out from the experience: the incertitude, the suffering, the conflicts (so often at the beginning of an ethical reflection) come from the experience. This experience can be shared as it is plural but unique for every carer and being ambivalent too (as it is a source of creativity and discouragement as well). This sharing will allow the subjectivity to come out in taking into account the suffering, the incertitude and the conflicts. The danger exists that the experience represents only a simple way of knowing without sharing. Then it would only help structure the own power of professional bodies. It would be richer if there is a collective ethical reflection allowing a better evaluation of situations in which the risk of an excess of medical and caring responsibility exists. Setting an ethical function in hospitals will fight this danger which comes from the reality of the institutional functioning.

THE INSTITUTIONAL FUNCTIONING

Tension between the" institué /instituant"

Medicine and hospitals are two institutions. They aim to assume a global social regulation, essential to the good working of any society; they contribute in their way to

> "the maintaining or renewing of the community's forces as they allow people to live, to love, to work, to change and perhaps to create the world according to their image"(Enriquez, 1996:62)

So an institution is first concerned by the human and social relations and by what those relations are sustained symbolically and imaginarily. What characterises the institution is to be centred on the

> "problem of alterity that means the acceptation of the other as a thinking autonomous subject by the social actors who have some affective and intellectual relations" (Enriquez, 1996 62-63).

So a real institution should wish and let the experience of subjective singularities. From particular situations to other particular situations the process of institutionalisation of an ethical reflection could then be set up on the experience of belonging to the world as a creative "instituant" capacity being part of what has already been "institué".

But it does not happen as serenely as that. The aims of the institution can be fulfilled only if a substantial agreement exists about these purposes. This agreement of course can be possible because it is laid by cultural and social norms which have been integrated by the carers. Nevertheless in spite of the institutional efforts to be peaceful and to create a consensus which is necessary the institution generates a certain violence, some tensions and anguishes which often are unbearable for the members. To be successful the institution must in fact protect itself from what the people put in common in it, it must also, in order to survive, settle

> "some defences against the undefined, pulsions, others, the unknown, the free speech, the thinking"(Enriquez, 1996:65).

This violence seems to be as substantial to the institution as the will to have a certain harmony in the social relations.

Harmony and violence are the two sides of the institution, between which any dynamic institution lives: (i) the "instituant" side of the creation, of the emergence of the new; (ii) the "institué" side of the consolidation of what has emerged as new.

This tension between "instituant" and "institué" is part of any institution. More precisely, the "instituant" acts as being able to question, to provoke a change of the "institué", which means a transformation of all the cultural, symbolic and imaginary evidences, stabilised by an institution. Any institution tends to privilege the "institué" dimension to the prejudice of the "instituant" dimension, maybe because setting up a new significant and structuring dimension of the human existence seems to be more complex than carrying on what has already been on and has been proved efficient.

The risk of "institué" alienation

The institutional reality shows this process of institutionalisation. The "institué" institution fixes the necessary stable functions to the social life and to the psychic life, in order to make the social exchange last with the risk of a stiffening of the institutional relations (terms). As far as the hospital is concerned, for example, the therapeutic aim of the institution always risks to be subordinated to the aims of an organisation which tends to consider only its specific functioning of organisation. The therapeutic process then risks to be taken by the bureaucratisation of the organisation: the "institué" supplants and reduces the "instituant" function of the institution to the detriment of patients of course, but also of the carers and the professional bodies of the caring institution. On that account the ethical function which is unambiguously on the side of the questioning risks to suffer from such a reality

In the dialectics "instituant"/"institué", the insistence on the "institué" time of the institution can be seen in the functioning of institutions by the elaboration of a culture adapted to each institution which intended to shape the attitudes of the members of an institution. A system of thinking and acting, of values and norms is worked out for the purpose of facilitating a collective creation. This system plays a significant role in the institutional life and guarantees the identity of the institution as a whole in the time. To give meaning to the practices and to the life of the members of an institution, symbols are also given (for example reference to mythical or real founders) which permit to gather the members for a project and to require honesty and involvement from them in the realisation of the institutional mission. When this culture plays its right role and is accepted correctly, it contributes to the good social and psychic functioning of the institution. Everyone can be occupied with his problems and make the institution live with a reasonably "instituant" dimension. This one can make the institution evolve and prevents it to be cut off from reality. On the contrary the demand of cohesion or the fear of what is outside the institution may ruin the good functioning of the institution. Then that culture invades the whole institutional space so much that it becomes the only possible reference. It does exist a risk to see people being caught in a pre-set meaning and being sunk (for the most involved, the most fragile, or the most credulous) in a suffering which can lead them to the alienation and to the psychic breaking down. This situation shows then the process of autonomy and the prevalence of the institution as for its members. It also shows the process of autonomy and the prevalence of the "institué" in the process of institutionalisation . Instead of creating some vital energy, the institution little by little will lead to a death-dealing process, for the members as well as for the society. In such a situation, the carers and the patients are in danger and some protecting mechanisms are worked out (absenteeism, and burn-out for the carers). The possibilities to elaborate an ethical reflection in such a context become very small.

TOWARD RELATIONS OF POWER WITHOUT ANY VIOLENCE

When the death-dealing process is recognised in time by the institution and when there is a will to fight it, this death-dealing process can however be a chance for an institutional revival. In order to have a psychic protection of the subjects of the institution and also a protection of the "instituant" capacity, it is then important to keep as big as possible

> "our capacity of thinking in the very moment when new institutional structures are searched and are tested" (Kaes, 1996:5).

This protection of a capacity of thinking (condition of setting up an ethical function) requires that the subjects of the hospital institution try their power in a responsible way, questioning the institution which employs them, and resisting to the institutional violence.

For if the use (a reasonable one if possible) of the power is necessary, it means that there is always an institutional violence.

> "Let us call violence any physical or psychic constraint, likely to carry along fear, move, unhappiness, suffering or death of a lively being; any act of intrusion which has the voluntary or involuntary effects of the dispossession of the other" (Héritier, 1996:17).

With that definition, we establish that violence turns out to be consubstantial with the human activity, that there cannot be a non-violence, meaning that it would be possible to eradicate, to put an end to this violence.

> "In an impassable way, violence is part of the forces which represent the real" (Defert, 1996:119).

It is not only due to "bad" persons with disastrous ideas and pernicious behaviours. These persons are always part of human groups (of an hospital for example) and violence inevitably belongs to the social and the politics.

An intrinsic relation combines violence to power even if power cannot be assimilated to

> "the category of the domination or even less to the negative category of the violence" (Defert, 1996:94).

In Foucault's perspective

> "the body is the target of an infinity of power tactics which give rise to intensification, desires, pleasures, identifications. The physical violence is only an extreme point among an infinity of rarely explored verbal and physical relations, of mechanisms which search, lash, investigate and transform" (Defert, 1996:96).

> "There is nothing like power or some power which would exist in bulk massively or in a diffuse concentrated or distributed state. There is power only exercised by some on others; power exists only in acting (...). It does not in itself give up a liberty (...). A relation of violence acts on a body, on things: it forces, bends, breaks, destroys; it has in itself all the possibilities; it has not another pole but the pole of passivity. If it meets a resistance, it has no other choice but the one of undertaking to reduce it . On the other hand a relation of power ties up with two elements which are absolutely necessary to be only a relation of power: the other on which the power is used must be recognised and held as subject of action and in front of that relation of power a number of responses, reactions effects and interventions must be opened " (M. Foucault, Dits et écrits , Paris, Gallimard, 1994, vol. IV, n° 306, p. 236; quoted by Defert, 1996:96).

We realise that if

> "any exercise of power cannot go without violence and consent (assent) they are not considered as the principles of power but they are the instruments of it " (Defert, 1996:96),

so no relation of power is condemned to the exercise of violence.

As we know that violence is present in a structuring way in the social relations, there is no reason why we should not find it again in the caring act which is a social relation. This violence can specially exist in the hospital context which in many ways lays down its institutional order – for example patients suffer from the rhythm of life, from the forbiddings which are not always justifiable, from the incomprehensible and unexplained therapeutic decision to the systematic giving sedative medicine, from the reduction to the symptom to the denial of the ill person. This violence which concerns each one as a human being turns out inevitably to be felt or acted by the carers as well as by the cared people. The situations in which the intervention of the professional bodies generates violence are not uncommon. In any case in the contemporary hospital, a process of breaking up the persons who enter, which ends at their reification, to their self effacement as subject, is at work.

We cannot only fight this violence by the punishment of the trouble-makers as this violence can be the act of each of us and also as this violence is the possible instrument in the relation of powers. On the other hand we can work out to set up relations of power which will be concrete and ridded of their violence:

> "these relations of power are movable, reversible and unstable (...) There can be relations of power only in so far as the subjects are free. If one of those was completely at the service of the other and was becoming (its property) its own object on which it could make use of an infinite and unlimited violence, there would not be any relation of power (...) That means that in relations of power there is inevitably a possibility of resisting (...) of resisting strongly of running away, using trickery, using strategies which would reverse the situation" (M. Foucault, op. cit., vol, IV, n° 356, p. 720; quoted by Defert, 1996:112).

So the permanent instability of the relations of power allows a questioning of this violence which is not fixed once for all on the same persons'side.

> "How to pull violence out of the darkness and the habit which make it nearly invisible? How to bring it out among the dullness of general mechanisms which makes it unavoidable and eventually bearable. The hidden violence can be defied to let it come out from the ruled moulds in which it is integrated " (M. Foucault, op .cit., vol. III, n° 191, p. 139; quoted by Defert, 1996: 113).

The development of the biomedical technoscience and its working out in the hospital institution represent a kind of knowledge, technology and power. Facing this, we must produce individualities who can not be passive, who can prove being resistant to this

> "confusion of violence, passions, hatreds, and revenges"; (M. Foucault, op. cit., vol. III, n° 272, p. 803; quoted by Defert, 1996:109).

Showing any resistance to violence does not only mean criticising the institution or, by putting forward law and justice, denouncing the violence of the values which are recognised by nearly everybody but which are so general as they are often used for the privilege of some only, for

> "rules are empty, violent and not finalised" (M.Foucault, op.cit., vol. II, n° 84, p. 145; quoted by Defert, 1996:103).

To resist to violence means to fight it, to identify its opponents (very often oneself) to fight for a victory, that is to say, for the recourse to a rule which would be finalised again so that

> "violence is done to violence " (M. Foucault, op. cit., vol. II, n° 84, p. 145; quoted by Defert, 1996:103).

There we come back to the aim of an ethical function in hospitals and to the ethical approach in its inevitably political component . For the ethical reflection allows to become aware of new aspects of reality: our own reality as a person who is ambivalent facing violence, the realities of other, of the institution, of the society and of the violence. This knowledge comes from the fight as it aims, in an ethical approach in any case, to critically evaluate the quality and the pertinence of the human action, having regard to the demands of the subject and to the constraints of the situations, in order to give back its sense of limits to the human practice, at the risk of falling into excessiveness and inhumanity especially the excessiveness and inhumanity (so human) of violence. The elaboration of an ethical judgment, intrinsically fragile, which is nothing but an objectivising knowledge, a scientific corpus brought to light in case of problems as a miracle remedy, is then a risky adventure full of uncertainties. It is necessary to oppose an interpretation (the one which comes out from the ethical evaluation as elucidation) to another interpretation which has been forced itself as violence and which must be reversed or at least changed step by step.

If ethics have a chance to be effective, it is due to the existence and the demand of subjects aiming to their autonomy, being able to keep distance and to question, in other words, being able to take charge of the demands of the ethical aim they are working on, to become completely open to that ethical aim and so to re-create ethics every time. An ethical function is only conceivable and possible only if professional bodies accept themselves as subjects and try from their practice to change the state of things. When these professional bodies are involved in such a process, it is then possible to break the silence enjoined on them by their powerlessness with regard to actions open to criticism which are sometimes shocking but which end to be unseen by them through habit. Adopting a critical detached attitude allows a time of reflection which can play a part of regulation in the situation of violence and a part of prevention against the violence which exists in the therapeutic taking charge. The carer's responsibility is stimulated. It is then possible to consider a debate about the different possible orientations of an action rather than to be satisfied with confrontations in urgency due to the threatening of some events.

Breaking the silence of the hospital not to eradicate the violence, but to canalise it, to temper it, to try to restore the dignity to subjects who are together, the carers as well as the cared, for the best as well as for the worst : this is the concrete task assigned by an ethical function, a process the gradual setting of which needs a slow transformation of the power relations within the hospital institution so that a subject itself can exist.

THE ETHICAL FUNCTION AS A CRITICAL ROLE

The power of imagination

The possibility of having an ethical reflection within an institution supposes that this reflection exists thanks to the act of a subject. In that way ethics is a praxis, an act which does not suppose the imitation of registered patterns. If such an approach has a normative dimension, this normativity does not come from positive contents but from the encouragement to the debate, to the critical questioning in comparison to the quality of the act which has been done or which is to be done. Briefly an ethical approach is an act without any pattern, more like an adventure or an exploration which aims to set up the conditions for a subject to exist through the human action. Such an adventure supposes the recognition of the "other", the desire that the other exists, thanks to joint responsibility practices among suffering subjects searching a sense to their caring practice. A reflective work, always being worked on, with those suffering subjects searching for a joint responsibility and for a sense to it is the position of ethics. Such a position sends back to the deep nature of medicine not only as a given service but also as a practice to the meaningful human aspects (illness, suffering, death), practice which wonders about the quality of life which has its pertinence only in the weaving of intersubjective and also significant ties.

Such a position is not so obvious in a hospital as a technique place. Setting up human signifying spaces (helping reflection in services for example) requests that the technological knowing how and the acting must not be only used on the technical part but to be significant as part of a more global debate taking into account the desire of the concerned subjects. In other words the doing and the saying must be sufficiently integrated to each other : if the doing goes without saying, it has no meaning and if the saying goes without doing it is banished from the questions of life. Such an integration cannot be successful without a recognition of the asymmetry existing within the institution between the technical speech, always present and powerful, and the human speech of ethics.

A clinical ethics practice centred on the narration of difficult situations by the concerned professional bodies allows such a recognition and allows to set up a caring subject as responsible for his acting willing to come to a finality (even an implicit one). A critical space of the technical rationality is being created then. Such a space inserts in this rationality a bit of history and life making a situation more complex in which the simple solutions (generally speaking about a suffering person) do not exist. The probably ineluctable asymmetry between the two types of speeches can be registered in a place where the human beings (here the professional bodies) could appropriate their history, could create some coexistence places, some places of signification, where the control of their destiny becomes possible, where the desire of another reality comes, where the suffering and the lack can be said at least. Through such a human exchange, a reflection about experience can be built up, an understanding of the experiences and the joint responsibilities can be worked out, a meaning is given from those experiences themselves. The practice of ethics must guarantee a sufficient safety, must guarantee that the weakness of the other will not be enjoyed and then this practice makes possible the building up of a signifying reality within the running reflection being produced.

Such places seem to be the indispensable condition to ask at least the question of changing the health institutions, question asked of course by the professional bodies of these institutions.

Such spaces, seen as places of experimentation and self training, will make possible the coming out of the subjectivity in the social field, essential dimension of the politics in which the ethical reflection has some ties. The invention and imagination, necessary to the institutional creation, could come out from those spaces more than from the present ethical committees, in order not to be satisfied to manage the ineluctable.

To produce another reality from the experience itself (knowledge and action, saying and doing mixed in an indissociated way) can be said being the political aspect of the finality of an ethical function in hospitals. Why shouldn't such places contribute to re-invent politics in today's world as individual responsibility and as scheme of communications between different social groups ?

The strategy for autonomous ethics

One question has to be asked: is helping the professional bodies to acquire the possibility to reflect upon the meaning of their acting a bit risky ? Encouraging them (those who agree) to begin to speak and to be conscious of the importance of taking the floor, helps the professional bodies to be ready to accept a potential breaking with the institution and its logic. They will be then able to say : «We are responsible and autonomous, or at least we try to be» even if this will be difficult as this logic has power through collective determinations. Now the carer feels powerless by those collective determinations (the norms of managing, the financing of the medical care system, the increasing specialization of caring roles and so the increasing division of the tasks….). Even if indeed there is a risk, knowing there is a space of debate will make the institution be aware that they cannot ask the professional bodies to be responsible and autonomous without giving them the means to be. Moreover, according to our hypothesis, there are still some forms of resistance from the subjects in the institution which are opposed to the forms of conditioning existing in the caring practice. These subjects , indeed fragile because lucid, would always like to bring in their situation a different logic from the one which seems to be forced on them ineluctably.

It is advisable to watch three very simple conditions to multiply the chances for a real autonomous ethical approach with regard to the institution , for an approach which does not supply the deficient functioning and institutional norms.

First, an autonomous ethical function must master its questions, it must be able to ask the right questions , those which are essential. If some questions are asked by the institution, the autonomous ethical function must be able to refuse or to reformulate them in a more pertinent way. Then, this function must listen to the subjects of the institution and so promote the taking the floor and the creation of spaces of debate in the institution. Finally, this function must be the acting of the subjects themselves in connection with their own questions and their own needs.

The legal existence for the ethics committees is a real chance to set up such an ethical function. Nevertheless such a legal framework in itself is not sufficient to guarantee the coming out of questions from the concerned subjects. Without the existence of these subjects able to ask their own questions and to have these questions discussed in the institution, such an autonomy in the ethical function could not be considered.

For an autonomous working, the legal framework must receive all the demands, the institutional ones as well as the personal ones; it must formulate questions in connection with those demands; it must work out the steps, judged necessary, for an effective answer to the question or an effective change of behaviour; it must evaluate the consequences a given answer will have in the institutional practice.

Considering a legal and autonomous framework as it has been described above, the question about the means assigned to this ethical function in hospitals becomes crucial : independent proceedings on the institutional level and autonomous proceedings on the conception level would not mean much if the means, specially the financial ones but also the human ones, were not given (which is the case for the moment in Belgium where no public financing is provided to carry on the functioning of ethical committees ordered by the law).

Once this frame has been set and financed, it must be made alive by the professional bodies of the institution. To exist, they must get out from the role the institution has defined for them; they must not be afraid of asking questions and showing their possible unsatisfaction. They can then manage the building up of an ethical questioning of the institution. But then a concrete problem is set : how to train the professional bodies of the institution involved in this questioning and how to give them an ethical support ?

Pierre Boitte, Centre d'Ethique Médicale, Université Catholique, Lille, France.

REFERENCES

Defert D. (1996), *La violence entre pouvoirs et interprétations dans les œuvres de Michel Foucault.* In *Héritier* F. (Séminaire de), *De la violence*, Paris : Editions *Odile* Jacob, pp. 89-121.

Enriquez E. (1996), *Le travail de mort dans les institutions.* in Kaes R., *L'institution et les institutions. Etudes psychanalytiques,* Paris : Dunod, pp. 62-94.

Héritier F. (1996), *Réflexions pour nourrir la réflexion.* in Héritier F. (Séminaire de), *De la violence,* Paris : Editions Odile Jacob, pp. 11-53.

Kaes R.(1996), *Réalité psychique et souffrance dans les institutions.* in Kaes R. *L'institution et les institutions. Etudes psychanalytiques,* Paris : Dunod, pp. 1-46.

NOTES

[1] This text was published in French with this title: "*Pour une fonction éthique à l'hôpital*" in Ethica Clinica, Revue francophone d'éthique des soins de santé, n. 15, Septembre 1999, 30-38 (Ethica Clinica : fih-w@openweb.be).

[2] About the Belgium experience, see « L'éthique en petits comités ? Comités d'éthique hospitaliers et fonction éthique à l'hôpital », *Ethica Clinica,* n.15 septembre, 1999 (Ethica Clinica : fih-w@openweb.be).

[3] In different issues of *Ethica Clinica,* a Frenchspeaking Quarterly Journal in Health Care Ethics, you will find out one or the other of the examples of the results of such an approach which specifity consists in centring the ethical reflection on the experience of carers.

CORRADO VIAFORA

THE ETHICAL FUNCTION IN THE HEALTH CARE INSTITUTIONS: CLINICAL ETHICS COMMITTEES

INTRODUCTION

Within the evolutionary frame of bioethics from a theoretical model based on "applied ethics" to a theoretical model based on integrating "internal morality" and "external morality", the Clinical ethics committees movement plays an important role.

According to D. Roy (D. Roy, 1987), the specific aim of the clinical bioethics is to establish the ethical reflection within those institutions in which the new biomedical knowledges and powers are produced. In this perspective the creation and development of the Clinical ethics committees assumes a significant value as the priviliged strategy for accomplishing the integration of the "internal" and "external" morality.

Since the epistemological and normative status of this kind of ethics committees is still open, an adeguate evaluation of this particular attempt to promote ethical function in health care institutions needs both an historical reconstruction and a theoretical reflection (P.Cattorini, 1988, F. Abel, 1993; C. Viafora, 1994).

CLINICAL ETHICS COMMITTEES: HISTORICAL BACKGROUND AND THEORETICAL DEBATE

The North American perspective: from "advocacy" paradigm to the "organizational" paradigm

Clinical ethics committees originated in the United States in the early 1980s in response to some "extraordinary" cases where there was a conflict between physicians and patients about the justification of clinical decisions. When hemodialysis became available (around 1960), Clinical ethics committees were formed to deal with the delicate issues rised by this scarce new resources. In 1976 an ethical conflict developed between physicians and the family of Karen Ann Quinlan, a patient in permanent coma. The conflict was taken to Court and ultimately to the Supreme Court of New Jersey. The Supreme Court decided in favour of the family's interest in removing her from the lifesaving support. In his written decision, the judge suggested that such dilemmas should be worked out in the *clinical setting* and not given to the courts for resolution. The judge, moreover, recommended that in

C. Viafora (Ed.), Clinical Bioethics. A Search for the Foundations, 181-192.

similar cases a Hospital Ethics Committee review relevant facts and make a recommendation to the principal decision makers: that is patients, families, physicians.

In these terms J. Drane (J. Drane, 1994, pg. 1-16), in its explanation of the "basic facts about health care ethics committees", presents the frame of the questions which can be worked out in the Clinical ethics committees:

> "Conflict and differences of opinion about what is ethically right are also possible today among the many medical professionals involved in patient treatment; medical specialists, nurses, social workers, hospital administrators. How can what is ethically right be determined in this new context? With so many different valid perspectives possible on what is right, will the courts become more and more involved in medical care decisions? Or will medical professionals themselves join with community representatives to develop ethical guidelines? Should a Do Not Resuscitation order be written on an elderly patient who may arrest? How should informed consent to a DNR order be handled? What procedures should be followed when a patient is incompetent and highpriced technological life-support are doubtfully or marginally beneficial? Who determining patient competency to make decisions? Under what conditions should treatment be withhold from a handicapped newborn? What role do family members have on decisions about an elderly patient?"

The conclusion of J. Drane is the following:

> "The Hospitals ethics commttee is the best possibility for considering different perspectives which are likely to exist on what is best, mediating among different medical care-givers, and finding socially acceptable ways for handling the above mentioned questions which are full of ethical complexities."

In the North American context an increased public support for Hospital ethics committees came from the President Commission. In two separate reports ("Making health care decisions", 1982; "Deciding to forego life-sustaining treatment", 1983) the President's Commission both recommended the formation of Hospital ethics committees and stipulated their functions. Realizing that ethical problems would be ongoing, it suggested that hospitals set up their own HECs to promote good decision-making trough this specific function: (i) education in ethics both for professionals and community; (ii) development of ethical guidelines and policies; (iii) case consultation both as retrospective case review and perspective case review.

The year following the proposal of the President's Commission, another extraordinary and hard case, the Baby Doe Case, went through every level of the New York State Court system and was presented to the U.S. Supreme Court.

More recommendations were subsequently issued by the federal government that hospital set up their own Hospital ethics committees, especially hospitals with Neonatal Intensive Care Units. Particularly relevant was the endorsement of the American Hospital Association which contributed to the establishment of these committees. In its guidelines published in 1984, the American Hospital Association (J. Wilson, 1986) advanced the following reason in favour of the Hospital ethics committees:

> "The growth of medical knowledge and the rapid expansion of medical capabilities and technology have generated unprecedented opportunities and challenges in the delivery of health care. At the same time, this growth and expansion have created increasingly complex ethical choices for physicians, health care professionals, patients, and the families of patients. Recent efforts to clarify biomedical ethical issues on the

> institutional level have focused on the use of hospital biomedical ethics committees. Such committees, sometimes called 'ethics committees', 'human values committees', 'medical-moral committees', 'committees', 'human values committees', 'medical-moral committees', or 'bioethics committees', hold promise for identifying the ethical implications of these problems and their possible resolutions, if they are established with a clearly defined purpose and an understanding of their capabilities and limitations".

Since this influent recommendation, the reference to the "capabilities" and "limitations" of the Hospital ethics committees will be esplicitly noted within the theoretical and epistemological debate.

Just the titles of some important articles can express the persistent ambiguity referred to the Hospital ethics committees: "Ethics Committees: promise or peril?" (R. McCormick, 1984); "Pediatrics ethics committees: ethical advisors or legal watchdogs?" (R. Weir, 1987)."Giving answer or raising questions? The problematic role of institutional ethics committees" (J. Fleetwood, 1989); "Ethics committees and social issues: potential and pitfall" (D. Callahan, 1992); "The Hospital Ethics Comittees: health care institutional conscience or white elephant?" (D. Blake 1992).

The core of the theoretical debate followed to the first generation of the Hospital ethics committees was focused on limitations of the mission of the Hospital ethics committees in order to improve their specific potentialities toward the *advocacy* of rights of individual patient or the *conflicts resolution* in extraordinary hard cases. Little attention has been given to expanding the role of Hospital ethics committees to include broader institutional ethics issue. Since the increasing complexity of the healthcare institutions, the classic model of HEC seems inadequate (T Spencer, 1997, D. C. Blake, 1999).

This concept is clearly asserted by N. Wenger (N. Wenger, 2000) in hits proposal of "the HEC model of the future, builded on deficiencies of the past". The new model is proposed essentially in response to the irruption of the issues of *organizational ethics,* arguably the most important ethical issues presently addressed by health care institutions. Among these issues are those related to managed care, including concerns both about the conflict between the obligations of the physicians and about the effect of rationing schemas on the care of patients.

A model for the next generation of healthcare ethics committees should be based on four ideas.

First, ethics committee should be "proactive". This should not simply "work", on reaction to problem cases presented to their as an introspective sanctuary within the hospital where hard cases are dissected to their core. Second, ethics committee should be organizationally integrated and not isolated. The celebrate independence of ethics committees often brings with it marginalization with an organization. Third, ethics committees should be held accountable by measurable outcomes and not simply by good intentions. Finally, ethics committees should be oriented by *institutional values,* and not simply by the *legal rights* of patients.

According to this model, the proposed next generation ethics committee will be an "agent for systemic change, formally linked to other key committees (such quality management and executive medical) and responsible for accomplishing specific goals with measurable outcomes or behaviors. This next generation healthcare ethics committee is presented as way to shift ethical dimensions within healthcare institutions from its traditional activities of discussions and education to organization and management.

The European perspective: between public debate and medical decision - making

Since the late eighties, Clinical ethics committees had a rapid increase also in many European countries (P. Riis, 1994, G. Lebeer, M.Moulin, 1999-2000). Compared to the United States, the rise of the number of Hospital ethics committees in Europe was rather late. Starting as research ethics review committees, their role and scope expanded in the late eighties. Characteristic to Hospital ethics committee, in most European countries is that they combine different functions: reviewing research protocols, case consultation, developing guidelines, institutional policies and recommendations, and providing a forum for discussion of clinical-ethical issues among health care professionals.

This type of ethics committee does not seem to be equally applied in all countries. In the absence of harmonized data, however, this is more an impression than a conclusion. According to recent studies, still in progress, it appears more appropriate for the European context to speak of the *clinical ethics function,* which may be expressed both by the specific hospital ethics committees and by other forms of debating and educating.

The general goal is to "stimulate the vitality of ethics in the healthcare institution". The means can be different, as heterogenous are the organization, membership, status and methods of hospital ethics committees. They differ not only between various countries, but also within the same country. For reviewing research protocols, guiding principles have been formulated in international codes. In some countries, legislation or national guidelines concerning research in human subjects have came into effect. The proper role and function of ethics committees regarding tasks others than review of research protocols is less clearly defined. Reporting the final discussion held in Bruxelles (25-27 May 2000) within the second workshop of the European concerted action devoted to "the clinical ethics activity within the hospital", Guy Lebeer and Marie Luce Delfosse (G. Lebeer, M. Moulin, 2000) summarised as follows the European experience about the hospital ethics committees:

> "On the one hand ethics committees are conceived as structures that are immediately operational to assist medical decision making, on the other hand they are seen as "fora" to debate clinical issues within the hospital itself. So on one hand we have ethics that are directly operational for the use and benefit of healthcare practictioners and on the other a critical ethics reflection capable of addressing a broad range of issues that are not predetermined and as such not necessarily suited for immediate practical concretizations".

The title of this final report is significatively "Clinical ethics committees: between public debate and medical decision-making".

As in United States also in Europe this new institutional creation and/or extended function raises a number of questions and a wide range of opinions. Although views on this subject reflect individual feelings, they also deeply mirror their culture and institutional context.

The *physicians* seem reluctant because the hospital ethics committees can undermine or weaken the patient-physician relationship or on the contrary favour the delegation of the responsibility. Perplexities raise also from *juridical* point of view: when fundamental rights are involved, the hospital ethics committees are inadequate.

In the countries, moreover, in which the contraposition between religious and secular tradition is persistent (the Italian context in particular) the hospital ethics committees must cope with additional difficulties.

From *secular* point of view there is the fear that the catholics tempt to introduce confessional elements in the public healthcare institutions.

From *religious* point of view the opposition to hospital ethics committees is expression of more general opposition in considering the consensus as instrument of building the ethical norms.

In a more analytical manner one can resume as follows the reasons "pro" and reasons "contra" the institution of the hospital ethics committees on the European context.

REASONS "PRO"

The Clinical ethics function undertaken by hospital committees offers a possibility for discussion from an ethical point of view a number of questions involving the patient which up to now were not discussed within the clinical contexts and were limitated to the level of the medical decisions. The opening of these questions to disciplines beyond the hospital as well as to non professionals would also enable "public area" to be created in the hospital, where clinical problems could be discussed.

The Clinical ethics committees can also offer the incomparable advantage of being closer to clinical experience and not constrained by the abstract generalities. For this the participation to work of an ethics committee may be a very significant educational experience, since the any genuine bioethical education requires both reciprocal interplay between general principles and particular clinical cases and a real integration between different professional sensitivities involved in the clinical contexts.

The Clinical ethics committees would promote a concrete communication between the health care institution and the community. In this perspective these new institutions can both help the our modern society to develop its capabilities in regulating the biomedical progress and to promote awareness of human rights and dignity when people need to use the health services, and to interprete and apply principles and norms coming from different sources: international recommendations, national and local laws, deontological code and guidelines.

REASONS "CONTRA"

The Clinical ethics committees appear one more administrative structure on a context that already suffers from bureaucracy. Having the committee play an advisory role would needlessly complicate the medical decision process.

Rather than *opening* the hospital to ethics, creating an "ad hoc" structure in this area runs the risk of *enclosing* ethics in a specific body of experts.

Lastly, another criticism is that these committees could provide the occasion for a professional ethos to be affirmed and legitimated, at a time when the means used by professionals to maintain their power are put increasingly into questions.

Clinical ethics committees: potentialities and ambiguities of a developing system

Hospital ethics committees can be established for various reasons and follow different trends, often implied ones, but which affect their working and decisions. The *potentialities* and *ambiguities* linked to the hospital ethics committees are very clearly outlined in one of the literature few studies to be carried out in depth by M. A. Parizeau and based on a complex empirical study on Quebec ethics committees. Some issues raised by the authors of this study are worth reading (M. H.Parizeau, 1995).

Does the institutionalisation of ethical function by promoting an organism specifically devoted to this function mean to create a starting-point for a beginning of a systematic debate, or does the same mean to circumscribe the moral debate by the typical bureaucratic attitude to delimit the questions and to delegate them to a specific body?

Does the constitution of ethics committees favours the autonomy of patients and defend their rights, or is it an expression of some sort of a resistance to the autonomy of patients by inviting the citizens to accept decisions made by experts?

Finally, are the ethics committees founded to help with decision making the physicians who have to cope with complex ethical problems, or, differently, are the committees founded to ensure people that the health decisions ethically relevant are not made by the physicians only?

According with the analysis of M. H. Parizeau, I think that a consistent approach to these questions involves the definition of the relationships between *ethical norms* and the other norms governing the health care institution, i.e., *professional* norms, *organizational* norms and *legal* norms.

Given that this relationship is the most relevant issue on the evolution of the ethical debate within health care institutions, this parameter will be a reference-point on a survey of how hospital ethics committees were promoted and are working.

I do not mean to perform a quantitative analysis, which would be very difficult considering the fact that the situation is rapidly evolving. I think that it would me more productive to sketch the patterns to which the different existing committees can be related.

My aim is to point out the tension, the ambiguity as well as the potentialities of this developing system.

Ethics committees based on a professional perspective

This model of ethics committee is promoted by a group of health professionals, mainly physicians working in peculiar clinical context, where the ethical problems raised by the new medical technologies are more likely to be a component of their professional practice. The increasing medical technologies inherent to the development of new diagnostic and therapeutic treatment can make the good intentions of medicine doubtful. Taking into account the technological powers of modern medicine, we are no longer sure that the "maximum" of the intervention always correspond to the "best" interest for the patient.

A further reason, more or less explicitly stated, is represented by the attempt to find the solution of the new bioethical problems in a professional perspective. The decision to establish ethics committees is the response to the perception that one's professional superiority is endangered: the risk that norms external to the professional body can determine the behaviours which need to be adopted.

The *potentialities* of this model are oriented: (i) to help the health professionals dealing with increasingly complex ethical problems; (ii) to establish a proper space in order to integrate the ethical sensitivities of different professionals involved in a clinical context.

The *ambiguity* of this model is essentially connected with the tendency of professional bodies to manage the new ethical problems in order to delimit the bioethical debate within the professional norms.

Ethics committees based on an organizational perspective

This model of hospital ethics committee is directly set up by the hospital administration. The reason leading to its institution is the awareness that ethical aspects become increasingly important in improving the quality of health care services. A further reason, more or less explicitly stated, is represented by the tendency of the institutions to "institutionally direct" the debate on ethical issues raised within the its clinical contexts. In other words: decisions of concern to health workers, along with "objective" criteria technically measurable and that can be assessed from "outside" according to specific standards of efficiency and competence, are increasingly comprehensive of "subjective" criteria appealing to ethical reasons and personal meanings. That's why administrators are willing to have recourse to mechanisms suited to directing greatly subjective decisions as decisions ethical in nature.

The functions of ethics committees set up in this perspective tend mainly to answer to the "good management" of the institution, and this involves potentialities and ambiguities.

The *potentialities* of this model are oriented: (i) to support the hospital administration with a proper counselling in order to give credibility to the institutional policies with ethical implications; (ii) to sensitive health care professionals to the ethical dimension of their work in order to involve them more directly into the institutional responsibility.

The *ambiguity* of this model is essentially connected with the tendency of the hospital administration to control the risk involved in the open ethical debate and consequently to delimit the bioethical debate within an institutional setting.

Ethics committees based in a public perspective

This model of ethics committee is marked by its reference to the public function of the health care institution. A *citizen*, who uses the health services, is the main subject of the attention of this committee, obviously without excluding the *health professionals* and the *hospital management*. When this perspective prevails, the committee's horizon becomes considerably wider. In this perspective the hospital ethics committees are considered as "examples of democracy". When this perspective prevails, it becomes clearer that institutional norms are not enough to manage the new ethical problems connected with the biomedical progress.

This model is grounded, on the one hand, on the belief that deontology centred on physicians' *duties* for the patients provides a specific competence to guide physicians' behaviours, but it is unable to lead with patients' *fundamental rights* in order to balance them, in case of conflict. On the other hand, there is an awareness that it is increasingly difficult to balance these rights due to the current moral pluralism.

The presuppositions inspiring this ethics committee are, substantively, the perception of the role played by human values at stake in the clinical contexts and the perception of the difficulty in finding shared ethical principles in the public area.

The *potentialities* of this model are oriented: (i) to make more transparent and comprehensive the ethical debate within health care institutions; (ii) to promote a real communication between the groups of professionals directly involved in the management of the biomedical progress and the community.

The *ambiguity* of this model is essentially connected with the difficulty for the hospital ethics committees to become "representative of civil society", the difficult in other words to become a symbol of democracy.

THE INSTITUTIONALISATION OF THE ETHICAL FUNCTION: PROCEDURAL, FOUNDATIONAL AND EDUCATIONAL CONDITIONS

M. H. Parizeau commenting on the results of the survey on the ethics committees established in Quebec hospitals, asserts that only the real assumption of their *specific ethical function* will be able to guarantee the credibility of this kind of ethics committees.

> "The analysis about the task of the clinical ethics committees in helping health workers stresses the strength and weakness of this structure which is emerging within hospitals. The aroused difficulties illustrate the need for the Committees to define their specificity and to get the instruments to guarantee this specificity".

M. H. Parizeau thinks this is the only way to prevent, on the one hand, the creation of useless duplicates and, on the other, the polarisation between the protection of the professional interests and that of the administrative management.

The clear reference, contained in this final warning, to a possible manipulation of ethics committees (i.e. the protection of professional interests and of the administrative management) arouse the most important ethical-foundational issue involved in the debate on the hospital ethics committees.

This is in brief the relation between the specifically *ethical* decision-making task of the Committee and the hospital *institutionalising norms*: the professional norms, the managerial norms, the legal norms.

In this perspective the most recurrent attitudes are two: the *first* is to think it possible to comprehend bioethical problems within the institutionalised rules. The *second* consists in regarding the health care as a mere technique void of any *internal morality*, consequently demanding the application of *external* general rules and principles.

Both attitudes are inadequate. The first, by thinking it possible to comprehend the solution of bioethical problems within the "professional" or "managerial" or "legal" norms is unable to give the wider "communicative" horizon, represented by the reference to the "man's reason", the specific ethics horizon, whose mention is decisive in the conflicts between the demands of the different institutionalised norms. The second attitude also is inadequate since, by pretending to place itself beyond any particular norm, it is unable to understand the morality "internal" to health care practices and institutions.

What are then the *conditions* enabling the ethics committee to work according to its specific perspective?

Procedural conditions

The first condition is that the composition and the institutional role of the ethics committee are guarantees of its independence. In this point of view, it is very important to pay attention to an adequate balance between the needs of a real independence and the needs of a close relationship with the professional and administrative management in order to guarantee a real impact of the ethics committees function (C. Byk, G. Memeteau, 1996).

The second condition is that the committees allow each opinion and point of view to be freely expressed. Debating ethical issues demands giving more importance to the arguments than to the formal status of their supporters. According to this perspective, ethics committees have to come with the difference between a *directive* attitude, which is a leading one in the medical field, and the *argumentative* attitude, which is typical for ethics (J. Welie, 1995).

The third condition is that debate must not to be avoided and potential conflicts which may arouse during the discussion must not be "neutralised", before a deep insight has been obtained. This might happen only if the committee's internal cohesion, or the will of obtaining consent are the prevailing forces directing its actions. In this perspective, the ethics committee must be particularly aware of the tendency that recurs frequently inside the "expert group": to reduce each problem to a technical one. This reduction is supposed to be successful in dealing with a problem by inserting it into a specific competence and in avoiding possible conflicts (S. Moscovici, W. Doise, 1992).

The fourth condition is that the committee must pay attention to the different perspectives present in the health care institutions, stressing the importance of all of them by a dialogue open to all view. By keeping this goal in mind, the committee must pay a special attention to the typical administrative "trend" to adopt procedures aiming at solving conflicts. These procedures are based on the firm belief that the "co-operation" between the professional groups can settle each disagreement. Yet it is important, that "co-operation" does not mean "combined consent" (J. Moreno, 1989).

Foundational conditions

These conditions go beyond the methodological and organizational level and make reference to the problem of the relationship between the committees' profile and their ethical model. It is obvious that in the context of a totally "objectivistic" ethical model committees would be *unnecessary*. On the other hand, in the context of a totally "subjectivistic" ethical model their institution would be *impossible*. Between these two opposite ethical models, both inadequate to support ethical committees, the better answer may be given by an ethical model which refers to two different levels (D. Gracia, 1994).

The first level is based on the ethical convictions characterised by a general consensus. The basic reference point of these convictions is the principle of human dignity (as it is expressed on significant documents of European culture i.e. *European Convention of Human Rights, Bioethics European Convention, Recommendation on the Patient's Rights in Europe*).

The second level is based on convictions, which refer to "how" to act according to the principle of human dignity in the new settings opened by the biomedical progress. This is the *specific level* of deliberating activity of the ethics committee. It is a level that requires both a logical and historical building process, which can develop only through participative and democratic procedures. According to this perspective, the ethics committee provides the setting in which this building process takes place.

Educational conditions

The nature of the *deliberative work* of the ethics committees requests an adequate education. The risk is that if a specific education on moral reasoning fail, others attitudes and interests will be prevail: the interests of the dominant professions; the delegation of the problems *ethical* in nature to the *technical* expertise; the influence of the personality appearing dominant in the group; the merely formal respect of the organizational rules; the delegation of the problems, *ethical* in nature to the *legal* or *deontological* norms or *managerial* rules.

The most important educational purposes are: (i) promoting an adequate integration among members of the committee in working together in the specific field of the ethical deliberation, by privileging "common educational moments"(P. Poletti, 1994). For this it is important a clear definition of the mission of the ethics committee and it is similarly important that a common perception of this mission exists within the members of the ethics committee; (ii) individuating the needs of the healthcare professionals, of all users of the services and of their families evincing suggestions from the typology of the proposed case; from the committee members contacts with the respective operative contexts and from the reporting of difficulties perceived by the common opinion, (iii) knowing the most significant reference texts for the committee's deliberative activity and giving an adequate and systematic attention to the bioethical public debate developed in the community, on the basis of the clinical cases particularly conflictual; (iv) developing a continous activity with the purpose of applicating the procedures concerning the specific functions of the ethics committee: ethical analysis of clinical cases, formulation of guidelines, direction of educational programs.

Corrado Viafora, Professor of Moral Philosophy and Bioethics,Faculty of Educational Sciences and Faculty of Medicine, University of Padua, Italy.

REFERENCES

Abel F. (1993), *Comites de Bioetica. Necessitad, estructura, y funciones,* in: "Folia Humanistica", Septembre-Octobre, pp 389-418.

Blake D.C., (1992), *The hospital ethics Committees. Health care's moral conscience or white elephant?*, in: "Hasting's Center Report", Jan / Feb, pp. 6-11.

Blake D.C., (1999), *Organizational ethics creating structural and cultural change in healthcare* organizations, in: "The Journal of Clinical ethics", 3, pp. 187-193.

Byk C., Memeteau G., (1996), *Le droit des comités d'éthique*, Editiones A. Lacassagne, Paris.

Callahan D., (1992), *Ethics Committees and social issues: patential and pitfall,* in: "Cambridge Quarterly of Healthcare Ethics", 18, pp. 5-10.

Cattorini P., (1988), *I Comitati di etica negli ospedali*, in: "Aggiornamenti sociali", 6, pp. 415-429.

Comitato Nazionale per la Bioetica, *I Comitati etici*, Presidenza del Consiglio dei Ministri, Dipartimento per l'Informazione e l'Editoria, Roma 1992.

Drane J., (1994), *Clinical Bioethics, Theory and practice in medical ethical decision making*, Sheed and Ward, Kansas City.

Fleetwood J., (1989), *Giving answers or raising questions? The problematic role of institutional ethics committees*, in: "Journal of Medical Ethics", 3, pp. 137-142.

Fletcher J.F., (1991), *The bioethics movement and hospital ethics committees*, in: "Maryland Law Review", 3, pp. 859-894.

Lebeer G., Moulin M., (2000) (eds), *Ethical function in Hospital Ethics Committees*, Biomed II, Working papers: workshop 1, Bruxelles, 3-5 June 1999, Université Libre de Bruxelles – Institut de Sociologie; working papers: workshop 2, Bruxelles 25-27 May.

Mc Carrick P., (1992), *Ethics Committees in Hospital*, Scope Note 3, "Kennedy Institute of Ethics Journal", 3, pp 285-306.

Mc Cormick R., (1984), *Ethics Committees: promise or peril?*, in: "Law Medicine and Health Care", pp. 150-155.

Moreno J., (1999), *Consensus, contract and committees*, in: "Journal of Medicine and Philosophy", 4, pp. 394-408.

Moreno J., (1991), *Institutional ethics committees: procede with caution,* in: "Maryland Law Review", 3 (1991), pp. 895-903

Moscovici S., Doise W., (1992), *Dissensions et Consenses. Une théorie générale des decisions collectives*, Paris Presses Universitaires de France.

Nelson W., Wlody G., (1997, *The evolving role of ethics advisory committees,* in: "HEC forum", 2, pp. 129-146.

Parizeau M.H., (1994), *Comité d'éthique,* in: *Les mots de la bioéthique: une vocabulaire encyclopédique*, ed. by G. Hottois – M.H. Parizeau, Bruxelles De Boeck Université.

Parizeau M.H. (1995), *Hôpital et éthique. Roles et défis des comitées d'éthique clinique,* Québec, Les Press de l'Université Laval.

Poletti P., (1994), *I comitati etici, strumento di formazione,* in C.Viafora (a cura di), *Comitati etici.*

Reich W., (1990), *Perché i Comitati di etica? Una valutazione dei diversi modelli*, in: S. Spinsanti (a cura di), *I Comitati di etica in ospedale,* pp. 15-37.

Riis P. (1994), *Medical ethics in the European Community*, in: "Journal of Medical Ethics", 1, pp. 7-12.

Roy D. (1987), *L'éthique clinique,* in *La responsabilité étique dans le developpement biomedical,* Cathalises, Louvain-La-Neuve, pp. 263-281.

Spencer T., (1997), *A new role of institutional ethics committees: organizational ethics*, in: "The Journal of Clinical ethics", 4, p. 372-376.

Viafora C., (1995), *Comitati etici. Una proposta bioetica per il mondo sanitario*, Fondazione Lanza. Gregoriana Libreria Editrice.

Viafora C., (2000), *Ethical function in hospital ethics committees: potentialities and ambiguites*, in: J. Glasa (ed), *Ethics Committees in central and eastern Europe*, Council of Europe, Strasbourg.

Viafora C., Gobber D., Drigo P., *The ethical function in the hospital: the experience of Bioethics Committee of the Pediatric Department of the University of Padua,* in: G. Lebeer, M. Moulin (eds), *Ethical* function *in hospital ethics committees*, pp. 67-74.

Weir R., (1987), *Infants ethics committees: ethical advisors or legal watchdogs?*, in: "Law Medicine and Health Care, 3, pp. 99-109.

Welie J., *La centralità della funzione formativa nell'esperienza dei Comitati etici. Riflessioni del* segretario *di un comitato etico europeo*, in: C. Viafora (a cura di), *Comitati Etici*, pp. 75-90.

Wenger N., (2000), *The healthcare ethics committee model of the future builded on deficiencies of the past*, in: "HEC forum", 12, pp. 33-38.

Wilson Ross J., (1986), *Handbook for hospital ethics committees*, American Hospital Association Publishing, Chicago.

GENERAL BIBLIOGRAPHY

Abel F., (1993), *Comites de Bioetica. Necessitad, estructura, y funciones*, in: "Folia Humanistica", Septembre-Octobre, pp 389-418.

Agazzi E., (1998), *La malattia come vissuto umano e come oggetto della medicina*, in "Kos", 15, pp. 18-21.

Agich G. J., (1993), *Autonomy in long term care,* Oxford University Press,

Angelini G., (1997), *Corpo proprio e forma morale*, in Bertuletti A. et Al., *L'io e il corpo*, Milano, pp. 205-237.

Aristotele (1984), *Nicomachean Ethics,* in *The complete works of Aristotle*, The Revised Oxford Translation ed. Jonathan Bames, New Jersey: Princeton University Press.

Autiero A., (1997), *Elaborazione sistematica dell'argomentare etico,* in *Verità e veracità,* Atti del XVI Congresso Nazionale, ATISM, Napoli , pp. 117-139.

Azzone G. F., (1998), *Medicine from art to science: the role of complexity and evolution,* IOS Press, Tokyo- Amsterdam.

Baertschi B., (1998) *Les méthodes de résolution des cas*, in "Bioethica Forum", 26, pp. 4-11.

Bauman Z., (1993), *Postmodern Ethics,* Oxford/Cambridge, Blackwell.

Bauman Z., (1994), *Alone again. Ethics after certainty*, Demos, London.

Bauman Z., (1995), *Life in fragments. Essays in postmodern morality*, Blackwell, Oxford.

Beauchamp T. L., and Childress, J. F., (2001), *Principles of biomedical ethics* (5th *ed.)*, Oxford University Press.

Benciolini P., Viafora C., (a cura di), *Il rapporto medico-paziente*, in Quaderni di Etica e Medicina, CIC Edizioni, Roma.

Bender L., (1992), *A feminist analysis of physician-assisted dying and voluntary active euthanasia*, Tennessee Law Review 59, pp. 519-546.

Benhabib S., (1996), *Situating the self: gender, community and postmodernism in contemporary* ethics, Polity Press, Princeton.

Beresford E., (1996), *Can phronesis save the life of medical ethics?,* in "Theoretical medicine" 17, 3, pp. 209-224.

Berti E., (1984), *La razionalità pratica tra scienza e filosofia*, in AA. VV, *Il valore. La filosofia pratica fra metafisica, scienza e politica*, Gregoriana Editrice, Padova, pp. 11-26.

Berti E., (1988), (ed.), *Tradizione e attualità della filosofia pratica*, Marietti, Genova.

Berti E., (1988), *La razionalità pratica: modelli e problemi*, Marietti, Genova.

Berti E., (1990), *Etica ermeneutica ed etica argomentativa*, in Casamassima F., Fabris M. (a cura di), *Filosofia ermeneutica tra critica e progetto*, IRRSAE, Bari, pp. 117-128.

Berti E., (1993), *Soggetti di responsabilità. Questioni di filosofia pratica*, Diabasis, Reggio Emilia.

Bishop M., Nichelle Ch., Darragh M., *Organizational ethics and health care: expanding bioethics to the Institutional Arena*, Scope Note 36, National Reference Center for Bioethics Literature, Georgetown University, pp. 189-208.

Blake D. C., (1992), *The hospital ethics Committees. Health care's moral conscience or white* elephant*?* in: "Hasting Center Report", Jan / Feb, pp. 6-11.

Blake D. C., (1999), *Organizational ethics creating structural and cultural change in healthcare organizations*, in: "The Journal of Clinical ethics", 3, pp. 187-193.

Bobbio N., (1990), *L'età dei diritti*, Einaudi, Torino.

Bohman J., (1996), *Public deliberation: pluralism, complexity, and democracy,* The MIT Press, Cambridge

Botti C., (2000), *Bioetica ed etica delle donne. Relazioni, affetti e potere*, Zadig, Milano.

Brody B., (ed.) (1988), *Moral theory and moral judgments in medical ethics*. Kluwer Academic Publ., Dordrecht.

Brody H., (1987), *Stories of sickness*, Yale University Press, New Haven London.

Brody H., (1992) *The healer's power,* Yale University Press, New Haven.

Brody H., Miller F.G., (1998), *The internal morality of medicine: explication and application to managed care*, in "Journal of Medicine and Philosophy", pp.384-410.

Brock D. W., (1993), *Life and death : philosophical essay in biomedical ethics*, Cambridge University Press, Cambridge.

Byk C., Memeteau G., (1996), *Le droit des comités d'étique*, Editiones A. Lacassagne, Paris.

Cadoré B., (1994), *L'expérience Bioéthique de la responsabilité*, Artel, Louvain-la-Neuve.

Cadoré B., (1995), *Éthique clinique et débat bioétique*, in "Le Journal de l'infermiére de neurologie et de neurochirurgie", 1-2.

Cadoré B., (1995), *La méthodologie en éthique clinique,* in "Forum Ethique. Quelle éthique pratique en médìcine? ", Editions E.C.C.E., Bruxelles, pp. 19-43.

Cadoré B., Van Gils P.Y., Englerty M., (1996), *Quell'éthique pratique en médecine?,* E. C. C. E., Bruxelles.

Cadoré B., (1997), *L'étique clinique comme philospophie contextuelle,* Fides ed. , Montreal

Callahan D., (1981), *Minimalist ethics. On the pacification of morality,* in Caplan A., Callahan D., (ed.), *Ethics in hard times,* Plenum Press, New York 1981, pp. 261-281.

Callahan D., (1987), *Setting Limits,* Simon and Schuster, New York.

Callahan D., (1990), *What kind of life. The limits of medical progress,* Simon and Schuster, New York.

Callahan D., (1996), *The goals of medicine: setting new priorities*, in "Hastings Center Report", 25, special pp. 1-26.

Callahan D., (1998), *False hopes*, Simon and Schuster, New York.

Callahan D., (2000), *Universalism and particularism*, in "Hastings Center Report", 30, 1, pp. 37-44.

Callahan S., (1988), *The role of emotion in ethical decision-making*, in "Hastings Center Report" 18, pp. 9-14.

Campbell C. S., (1996), *Bioethics and the spirit of secularism,* in E.E. Shelp (ed.), *Secular bioethics in theological perspectives,* Kluwer Academic Publ., Dordrecht, pp. 3-18.

Carson R. A., (1990), *Interpretive bioethics: the way of discernment*, in "Theoretical Medicine", 11, pp. 51-59.

Casalone C., (2000), *Il tempo del medico e il tempo del paziente: un nuovo sguardo sulla relazione medico-paziente*, in Benciolini P., Viafora C., (eds), *Il rapporto medico-paziente*, pp. 42.51.

Casati S., (2003), *La deliberazione etica e la pratica clinica*, Edizioni Medico Scientifiche, Torino.

Cattorini P., (1996), *Bioetica. Metodo ed elementi di base per affrontare problemi clinici*, Masson, Milano.

Cattorini P., (1996), *I molti corpi del rapporto terapeutico. Per un'epistemologia della complessità in medicina*, in Cattorini P., Mordacci R., Reichlin M., (a cura di), *Introduzione allo studio della bioetica*, Milano, pp. 205-229.

Cattorini P., (1998), *Applicare la norma, interpretare il desiderio: il ruolo della bioetica clinica*, in "Kos", 15, pp. 60-64.

Cattorini P., (1999), *Applicare o interpretare? Il ruolo della bioetica clinica*, in "Rivista di teologia morale", 121, pp. 83-101.

Cattorini P., (2001), *Clinical bioethics. Identity, role, aims*, in "Medicina nei secoli. Arte e scienza", 13, pp. 187-197.

Cavicchi I., (2000), *La medicina della scelta*, Bollati Boringhieri, Torino.

Chadwick, R. (ed.), *Encyclopedia of applied ethics*, Academic Press, San Diego.

Charvet J., (1995), *The idea of an ethical community,* Cornell University Press, Ithaca, NY.

Childress J., (1997), *Practical reasoning in bioethics*, Indian University Press.

Clouser K. D., Gert B., (1990), *A critique of principlism*, in "The Journal of Medicine and Philosophy" 15, pp. 219-236.

Comitato Nazionale per la Bioetica, (1992) *I Comitati etici*, Presidenza del Consiglio dei Ministri, Dipartimento per l'Informazione e l'Editoria, Roma.

Comitato Nazionale per la Bioetica, (2001), *Scopi, limiti e rischi della medicina*, Presidenza del Consiglio dei Ministri, Dipartimento per l'Informazione e l'Editoria, Roma.

Daniel S.L., (1986), *The patient as text: A model of clinical hermeneutics*, in "Theoretical Medicine" 7, pp. 195-210.

Daniels N., (1996) *Wide reflective equilibrium in practice*, in Sumner L. W., Boyle J. (eds.), *Philosophical perspective on bioethics*, University of Toronto Press, Toronto.

Decision making methodology in bioethics (1991), 12, 4, (edited by) Erde E.

De Renzo E., (1997), *A feminist model for clinical ethics consultation: increasing attention to context and narrative*, in "Hec Forum", 3, pp. 212-227.

DeGrazia D., (1992) *Moving forward in bioethical theory: theories, cases and specified principlism*, in "Journal of Medicine and Philosophy", 17, pp.511-539.

DeGrazia D., Beauchamp T. L., (2001), *Philosophical foundations and philosophical methods*, in Sulmasy D., Sugarman J. (eds.), *Methods in medical ethics*, Georgetown University Press, Washington D. C..
Dell'Oro R., (1995), *Antropologia ed etica. Oltre la bioetica nordamericana*, in "Rivista di teologia morale", 106, pp. 96-114.
Dell'Oro R., (2000), *Antropologia dell'intimità e segreto professionale: il problema del controllo delle informazioni mediche*, in Benciolini P., Viafora C., (a cura di), *Il rapporto medico paziente.*
Devettere R., (2000), *Practical decision making in health care ethics*, 2nd ed., Georgetown University Press, Washington.
Dickenson D., Fulbord B., Murray T., (eds.), (2002), *Healthcare ethics and human values. An introductory text with readings and case studies*, Blackwell, Oxford.
Donagan A., (1993), *Moral dilemmas, genuine and spurios: a comparative anathomy*, in "Ethics", 104, pp. 7-22.
Doucet H., Larouche J. M., Melchin K., (2001), *Ethical deliberation in multiprofessional healthcare teams*, University of Ottawa Press, Ottawa.
Drane J. F., (1988), *Becoming a good doctor: The place of virtue and character in medical ethics*. Sheed and Ward, Kansas City.
Drane J. F., (1998), *Clinical bioethics. Theory and practice in medical-ethical decision making*, Sheed and Ward, Kansas City.
DuBose E. R., Hamel R., O'Connell, L.J. (eds.): (1994), *A matter of principles? Ferment in U.S. bioethics*, Trinity Press International, Valley Forge (Pennsylvania).
Dworkin R., (1977), *Taking rights seriously*, Harvard University Press, Cambridge.
Dworkin R. (1988), *The foundations of liberal equality,* in G. Petersen (ed.) "The Tanner Lectures on Human Values".
Dworkin R. (1993), *Life's dominion*, Cambridge University Press, Cambridge.
Ellos W. J., (1998), *Some narrative methodologies for clinical ethics*, in "Cambridge Quarterly of Healthcare Ethics", 7, pp. 315-322.
Emanuel E. J., (1991), *The ends of human life: medical ethics in a liberal polity,* Harvard University Press, Cambridge.
Emanuel E. J., Emanuel L. L., (1992), *Four models of the physician-patinet relationship*, in JAMA, 267, 2221-2226.
Engelhardt H. T., (1991) *Bioethics and secular humanism*, Trinity Press International, Philadelphia.
Engelhardt H. T., Wildes K., (1994) *The four principles of health care ethics and post-modernity*, in Gillon R. (ed.), *Principles of health care ethics*, Wiley, London.
Engelhardt T., (1996), *The foundations of bioethics,* Oxford University Press, New York.
Engelhardt Von D., (1989), *Ethik in Altag der Medizin. Spektrum der medizinischen Disciplinen,* Springer Verlag, Harberg.
Ethics consultants and ethics consultations, (1993), in "Cambridge Quarterly of Healthcare Ethics", 2.

Evans M., (ed.), (1998), *Critical reflection on medical ethics,* JAI Press, Stamford (Conn.).

Fantini B., (1995), *La constitution du savoir médical moderne et le caractère spécifique des maladies*, in “Cahier médico-sociaux”, 39, pp. 15-23.

Federspil G., (1991), *La malattia come evento biologico*, in id. et al., *La malattia e la filosofia della medicina*, Torino, pp. 5-26.

Federspil G., Vettor R., (1994), *Pensiero clinico e metodo clinico*, in “L’arco di Giano”, 2, pp. 101-119.

Finnis J., (1980), *Natural law and natural rights*, Oxford University Press, Oxford.

Finnis J., (1984), *Practical reasoning, human goods and the end of man*, in “Proceedings of the American Catholic Philosophical Association”, pp. 23 -36.

Fleetwood J., (1989), *Giving answers or raising questions? The problematic role of institutional ethics committees*, in: “Journal of Medical Ethics”, 3, pp. 137-142.

Fletcher J. C., (1995), *Clinical bioethics at NIH: History and a new vision*, in “Kennedy Institute of Ethics Journal”, 4, pp. 355-364.

Fletcher J. C., Brody H., (1995), *Clinical ethics-I. Elements and methodologies*, in Reich W. T. (ed), *Encyclopaedia of bioethics*, The Free Press, New York, pp. 309-404.

Fletcher J. C., Siegler M., (1996), *What are the goals of ethics consultation? A consensus statement*, in “The Journal of Clinical Ethics”, 7, pp.122-126.

Fletcher J. F., (1991), *The bioethics movement and hospital ethics committees*, in: “Maryland Law Review”, 3, pp. 859-894.

Foot P., (1978), *Practical reasoning,* in Reich W. T., (ed), *Encyclopedia of bioethics*, I ed., pp. 450-454.

Foucault M., (1975), *The birth of the clinic: an archeology of medical perception*, New York, Vintage Books.

Fox D. M., (1979), *The segregation of medical ethics: a problem on modern intellectual history, in* "The Journal of Medicine and Philosophy", pp. 81-97.

Fox R. C., (1989), *The sociology of medicine. A participant observer's view*. Prentice-Hall, Englewood Cliffs, N.J.

Frankena W. K., (1973), *Ethics*, Prentice-Hall, Englewood Cliffs.

Fulford K.W.M., Gillett G.R., Soskice J.M. (eds.): *Medicine and moral reasoning*. Cambridge University Press, Cambridge, U.K.

Gadamer H. G., (1993), *Wahrheit und method*, Mohr.

Gadamer H. G., (1993), *Uber die Verborgenheit der Gesundheit,* Suhrkamp Verlag, Frankfurt am Mein.

Gadamer H., (1999), *Vernuft in zeitalter der wissenschaft*, Suhrkamp Verlag, 1976.

Galvagni L., Cattorini P., Tavani M., (1997), *La consulenza etica nelle strutture sanitarie: esperienze e modelli a confronto*, in “Medicina Legale, Quaderni Camerti”, 3, pp. 539-571.

Gaylin W., Kass L., Pellegrino, E. D., Siegler M., (1988), *Doctor must not kill*, in “Journal of the American Medical Association”, 259, pp. 2139-2140.

Gillon R., (1986), *Philosophical medical ethics,* John Wiley & Sons, Chichester.

Gillon R., (ed.), (1994), *Principles of health care ethics*. John Wiley & Sons, Chichester.

Graber G. C., Thomasma D. C., (1989), *Theory and practice in medical ethics*, Continuum, New York.

Gracia D., (1989), *Fundamentos de bioética,* Eudema, Madrid 1989.

Gracia D., (1991), *Bioetica clinica*, in Spinsanti S., (ed), *Antropologia medica*, La Nuova Italia Scientifica, Firenze, pp. 43-69.

Gracia D., (1991), *Procedimientos de decisiòn en ética clinica,* Eudema, Madrid.

Gracia D., (1999), *History of medical ethics*, in ten Have, H.A.M.J. and Gordijn, B. (eds.), *Bioethics in a european perspective*, pp. 1-50.

Gracia D., (1999), *What kind of values? A historical perspectives on ends of medicine*, in Hanson M., Callahan D. (eds), *The goals of medicine: the forgotten issue in healthcare reform*, Georgetown University Press, Washington D.C., pp. 88-100.

Gracia D., (2000), *Los fines de la medicina en el umbral del siglo XXI*, in *La bioética, horizonte de posibilidad,* Asociaciòn de Bioética fundamental y clinica. 3° Congreso Nacional, La Corunya 4-6 Marzo 1999, Madrid, pp. 55-68.

Gracia D., (2001), *La deliberaciòn moral: el método de la ética clìnica*, in "Med. Clin", 117, pp. 18-23.

Gracia D., (2001), *Moral deliberation: the role of methodology in clinical ethics*, in "Medicine, healthcare and philosophy", IV, 2, pp. 223-232.

Gray J. A., (1997), *Evidence based health care. How to make health policy and management decisions,* Churchill Livingstone, Edinburgh.

Grmek M. D., (1994), *Le antinomie del pensiero medico*, in "L'arco di Giano", 2, pp. 33-40.

Gutmann A., Thompson D., (1997), *Deliberating about Bioethics*, in "Hastings Center Report", 27(3), pp.38-41.

Habermas J., (1990), M*oral consciousness and communicative action,* The MIT Press, Cambridge.

Habermas J., (1993), *Justification and application: remarks on discourse ethics*, Polity, Oxford.

Habermas J., (1996), *Between facts and norms: contributions to a discourse theory of law and democracy,* The MIT Press, Cambridge.

Hanson M. J., Callahan D., (1999), *The goals of medicine: the forgotten issues in health care reform*, Georgetown University Press, Washington D.C..

Hare R.M., (1993), *Essays on bioethics*, Clarendon, Oxford.

Hare R.M., (1994), *Methods in bioethics: some defective proposal*, in "Monash bioethics reviews", pp.34-47.

Harre R., Gillett, G., (1994), *The deliberative mind,* Sage, London.

Harris J., (1985), *The value of life*, Routledge and Kegan, London.

Harris J., (1992), *Wonderwoman and superman: the ethics of human biotechnology*, Oxford University Press, Oxford.

Have H. Ten., (1995), *The anthropological tradition in the philosophy of medicine*, in "Theoretical Medicine", 16, pp. 3-14.

Have H. Ten, (1995), *Approcci europei all'etica nella medicina clinica*, in, Viafora, C. (ed.), *Comitati etici. Una proposta bioetica per il mondo sanitario*. Fondazione Lanza-Gregoriana ed., Padova.

Have H. Ten, (1998), *Health care and the human body*, in "Medicine, Health Care and Philosophy", 1(2): 103-105.

Have H. Ten, (1998), *Images of man in philosophy of medicine*, in Evans, M. (ed.), *Critical reflection on medical ethics*, JAI Press, Stamford (Conn.), p.173-193.

Have H. Ten, Kimsma, G., Spicker, S.F., (eds.), (1990), *The growth of medical knowledge,* Kluwer Acad. Publ., Dordrecht.

Have H. Ten, Lelie, A. (1988), *Medical ethics research between theory and practice*, in "Theoretical Medicine and Bioethics", 19, pp. 263-276.

Have H. Ten., (1997), *From synthesis and system to morals and procedure - the development of philosophy of medicine,* in, Carson, R.A. and Burns, C.R. (eds.): *The philosophy of medicine and bioethics. Retrospective and critical appraisal*, Kluwer Academic Publishers, Dordrecht.

Hottois G., Parizeau M.H., (1994), *Les mots de la bioéthique : un vocabulaire encyclopedique*, De Boeck Université, Bruxelles.

Illich I., (1976), *Limits to medicine. Medical nemesis: the expropriation of health*, Boyars Publishers, London.

Ingrosso M., (1994), *Ecologia sociale e salute. Scenari e concezioni del benessere nella società complessa*, Milano.

Ingrosso M., (1994), *La salute come costruzione sociale. Teorie, pratiche, politiche*, Milano, pp. 82-118.

Jaspers K., (1986), *Der Arzt im technischen Zeitalter*, R. Piper, Munchen,

Jensen U. J., (1987), *Practice and progress. A theory for the modern health-care system*. Blackwell Scientific Publications, Oxford.

Jensen U. J., (1989), *From good medical practice to best medical practice*, in "International Journal of Health Planning and Management", 4, pp. 167-80.

Jensen U. J., Mooney, G., (eds.), (1990), *Changing values in medical and health care decision making,* John Wiley & Sons, Chichester.

Jonas H., (1979), *Das Prinzip Verantwortung*, Insel Verlag, Frankfurt am Mein.

Jonas H., (1985), *Technik, Medizin und Ethik. Zur Prazis des Prinzips Veranwortung,* Insel Verlag, Frankfurt am Mein.

Jonsen A. R., (1990), *Case analysis in clinical ethics*, in "Journal of Clinical Ethics", 1, pp. 63-65.

Jonsen A. R., (1990), *Practice versus theory*, in "Hastings Center Report", 20, pp. 32-34.

Jonsen A. R., Siegler M., Winslade W., (1998), *Clinical ethics: a practical approach to ethical decisions in clinical medicine*, McGraw- Hill, New York.

Jonsen A., (1998), *The birth of bioethics*, Oxford University Press, New York.

Jonsen A.R., Toulmin S., (1988), *The abuse of casuistry,* University of California Press, Berkeley, CA.

Kaes R., (ed.), (1996), *L'institution et les institutions. Etudes psychanalytiques,* Dunod, Paris.

Kanoti G. A., Youngner S. J., (1995), *Clinical Ethics-II. Clinical ethics consultant*, in Reich W. T. (ed.), *Encyclopedia of bioethics*, The Free Press, New York, pp. 404-409.

Kant I., (1785), Grundlegung der *Metaphysik der Sitten*

Kant I., (1788), *Kritik der Prakthischen Vernunft*
Kant I., (1797), *Metaphysik der Sitten.*
Kass L. R., (1990), *Practicing ethics: where's the action?,* "Hastings Center Report", 20: 5-12.
Kass L. R., (1975/1981) *Regarding the end of medicine and the pursuit of h*ealth, in "The Public Interest", 40, pp. 11-42.
Katz J., (1984), *The silent world of doctor and patient,* Free Press, New York.
Kekes J., (1997), *Moral wisdom and good lives,* Cornell University Press, Ithaca, NY.
Kettner M., (1999), *Discourse ethics: a novel approach to moral decision making*, in "International Journal of Bioethics", 10, 3, pp.29-36.
Kornblith H., (1993), *Inductive inference and its natural ground. An essay in naturalistic epistemology,* MIT Press, Cambridge.
Kuczewski M. G., (1997), *Fragmentation and consensus. Communitarian and casuist bioethics*. Georgetown University Press, Washington, D.C.
Kuhn T., (1970), *The Structure of Scientific Revolution,* 2nd ed. , University of Chicago Press, Chicago.
Kuhse H., Singer P., (1998), (eds) *A companion to bioethics*, Blackwell, Oxford.
La Puma J., Toulmin S. E., (1989), *Ethics consultants and ethics committees*, in "Archives of Internal Medicine", 149, pp. 1109-1112.
La Puma J., Schiedermayer D. L., (1991), *Ethics consultation: skills, roles, and training*, in "Annals of Internal Medicine", 114, pp. 155-160.
La Puma J., Schiedermayer D. L., (1994), *Ethics consultation. A practical guide*, Jones and Burtlett Publishers, Boston-London.
Ladd J., (1983), *The internal morality of medicine: An essential dimension of the patient-physician relationship,* in: Shelp, E.E. (ed.), *The clinical encounter*, D. Reidel Publishing Company, Dordrecht, pp. 209-231.
Ladrière J., (1977), *Les enjeux de la rationalité. Le défi de la science et de la technologie aux cultures*, Paris.
Ladrière J., (1997), *L'interpretation dans le jugement éthique*, in "Le Supplement", 202, pp. 143-172.
Ladrière J., (1997), *L'éthique dans l'univers de la rationalité*, Artel, Namur.
Lamau M. L., (1991), *Créativité éthique dans le champ de la bio-médecine*, in "Recherches de sciences religieuses", 12, pp.141-149.
Lebeer G., (ed), *Ethical function in hospital ethics committees*, JOS Press, Amsterdam 2002.
Leder D., (1988), *The hermeneutic role of the consultation-liaison psychiatrist.* "Journal of Medicine and Philosophy", 13, pp. 367-378.
Leder D., (1990), *Clinical interpretation: the hermeneutics of medicine*, in "Theoretical Medicine", 11, p. 13.
Leder D., (1994), *Toward a hermeneutical bioethics*, in: DuBose, E.R., Hamel, R. and O'Connell, L.J. (eds.), (1994), *A matter of principles. Ferment in U.S. bioethics*, Trinity Press International, Valley Forge (Pennsylvania), pp. 240-259.
Lindemann N. H. and J., (1995), *The patient in the family*, Routledge, New York.

MacIntyre A., (1981), *After Virtue: A Study in Moral Theory.* Univerity of Notre Dame Press, Notre Dame.

MacIntyre A., (1988), *Whose justice, whic rationality?*, in University of Notre Dame Press, Notre Dame.

MacIntyre A., (1990), *Three Rival Versions of Moral Enquiry,* Notre Dame, In: University of Notre Dame Press.

Macklin R., (1999), *Against relativism: cultural diversity and the search for ethical universals in medicine,* Oxford University Press, New York.

MacLean A., (1993), *The elimination of morality. Reflections on utilitarianism and bioethics,* Routledge, New York.

Malherbe J. F., Dethier, L., Ravez, L., (1996), *La fonction éthique dans le monde hospitalie*r, in "Ethica Clinica", 1, pp. 4-13.

Malherbe J. F., (1987), *Pour une éthique de la médicine*, Larousse, Paris.

Mc Carrick P., (1992), *Ethics Committees in Hospital,* in "Scope Note 3", in "Kennedy Institute of Ethics Journal", 3, pp 285-306.

McGee G., (1996*), Phronesis in Clinical Ethics,* in "Theoretical Medicine" 17 (4), pp. 317-328.

Methodology in medical ethics decision making, (1994), 15, 1, (edited by) Erde E.

Miller F., Brody H., (1995), *Professional integrity and the physician –assisted death,* in "Hastings Center Report", 25, pp. 8-17.

Momeyer R., (1995), *Does physician-assisted suicide violate the integrity of medicine?,* in "Journal of Medicine and Philosophy, 20, pp. 13-24.

Monagle J., (1998), *Health care ethics: critical issues for the 21st century*, Aspen Publishers, Gatherburg (MD).

Montague B., (1996), *The quest for moral foundations: an introduction to ethics.* Georgetown University Press, Washington, D.C..

Mordacci R., (2003), *Una introduzione alle teorie morali. Confronto con la bioetica,* Feltrinelli, Milano.

Moreno J., (1991), *Institutional ethics committees: procede with caution,* in: "Maryland Law Review", 3, pp. 895-903

Moreno J., (1999), *Consensus, contract and committees,* in: "Journal of Medicine and Philosophy", 4, pp. 394-408.

Mori M., (1990), *La bioetica: la risposta della cultura contemporanea alle questioni morali relative alla vita,* in Viano C. A., (a cura di), *Teorie etiche contemporanee,* Bollati Boringhieri, Torino, pp. 186-224.

Morreim E. H., (ed), (1994), *Bioethics, expertise, and the courts.* In "The Journal of Medicine and Philosophy", 22 (4), pp. 403.

Moscovici S., Doise W., (1992), *Dissensions et Consenses. Une théorie générale des decisions collectives*, Presses Universitaires de France, Paris.

Murray T.H., (1994), *Medical ethics, moral philosophy and moral tradition.* In: Fulford, K.W.M., Gillett, G.R. and Soskice, J.M. (eds.), *Medicine and moral reasoning,* Cambridge University Press, Cambridge, pp. 91-105.

Natoli S., (1985), *La costituzione dello sguardo medico: dal gesto terapeutico alla scientificità della* medicina, in "Nuovi Saggi di Medicina e Scienze Umane", Milano, Istituto scientifico San Raffaele, pp. 31- 70.

Nelson H. L., (1997), *Stories and their limits: narrative approches to bioethics*, Routledge, New York-London.
Nelson W., Wlody G., (1997), *The evolving role of ethics advisory committees,* in: "HEC forum", 2, pp. 129-146.
Newton A.Z., (1995), *Narrative ethics*. Harvard University Press, Cambridge (Mass.)
Nielsen K., (1997), *Naturalism without foundations,* Prometheus Books, Amherst, NY.
Nordenfelt L., (1996), *On medicine and other species of health enhancement*, in L. Nordenfelt, P. Tengland (eds.), *The goals and limits of medicine,* Almqvist & Wiksell International, Stockholm.
Nordin I., (1999), *The limits of medical practice*, "Theoretical Medicine and Bioethics", 2, pp. 105-123.
O'Neill O., (2001), *Practical priciples and practical judgment*, in "Hastings Center Report", 31, 4, pp. 15-23.
Palazzani L., (1992), *Bioetica dei principi e bioetica delle virtù: il dibattito attuale negli Stati Uniti*, in "Medicina e Morale", 1, pp. 58-85.
Parizeau M. H., (1989), *Bioéthique: méthodes et fondements*, in Cahiers scientifiques de l'ACFAS, 66, pp. 3 – 16.
Parizeau M. H., (1992), *Les fondaments de la bioéthique*, De Boeck-Wesmael, Bruxelles.
Parizeau M.H., (1995), *Hôpital et éthique. Roles et défis des comitées d'éthique clinique,* Les Presses de l'Université Laval, Québec.
Parker M., (1995), *The growth of understanding*. Aldershot, Avebury.
Parker M., (1999), *Ethics and community in the health care professions*, London, Routledge.
Parodi A., (2002), *Storie della medicina*, Edizioni di comunità, Torino.
Paul C., (2000), *Internal and external morality of medicine: lesson from New Zeland*, in "British Medical Journal", 320, pp. 499-503.
Pellegrino E., Thomasma D., (1981), *A philosophical basis of medical practice: toward a philosophy and ethic of the healing profession*, Oxford University Press, New York.
Pellegrino E. D., (1983), *The healing relationship: the architectonics of clinical medicine in the clinical* encounter, in Shelp E. (ed.), *The clinical encounter: the moral fabric of the physician-patient relationship*, D. Reidel, Dordrecht, pp. 153-172.
Pellegrino E. D., Thomasma D. C., (1988), *For the patient's good: the restoration of beneficence in health care*, Oxford University Press, New York.
Pellegrino E. D., Siegler M., Singer P. A., (1991), *Future directions in clinical ethics*, in "The Journal of Clinical Ethics", 1, pp. 5-9.
Pellegrino E., Thomasma D., (1993), *The virtues in medical practice*, Oxford University Press, Oxford.
Pellegrino E. D., (1993), *The metamorphosis of medical ethics: a 30 years retrospective,* in "JAMA", 9, pp. 1158-1162.

Pellegrino E. D., (1994), *The four principles and the doctor patient relationship: the need for a better linkage*, in Gillon R. (ed.), *Principles of health care ethics*, J.Wiley, New York, pp. 353-367.

Pellegrino E. D., (1998), *What the philosophy of medicine is*, in "Theoretical Medicine and Bioethics", 19, pp. 315-336.

Pellegrino E. D., (1999), *The goals and ends of medicine: how are they to be defined?*, in Hanson M., Callahan D., (eds.), *The goals of medicine*, pp. 55-68.

Pellegrino E. D., (2001), *Phylosophy of medicine: should it be teleologically or socially construed?*, in "Kennedy Instituete Journal of Ethics", 11, pp. 169-180.

Pence G. E., (1995), *Classic cases in medical ethics*, McGraw-Hill, New York.

Poirier S., Brauner D.J., (1988), *Ethics and the daily language of medical discourse*, in "Hastings Center Report", 18, pp. 5-9.

Potter R., (1996), *From clinical ethics to organisational ethics: the second stage of the evolution of bioethics*, in "Bioethics Forum", 12, pp. 3-12.

Prodi G., (1981), *Salute/malattia*, in "Enciclopedia"XII, Torino, pp. 394-427.

Rawls J., (1971), *A theory of justice*, Harvard University Press, Cambridge.

Reich W., (1993), *Alle origini dell'etica medica: mito del contratto o mito di cura?* in Cattorini, B., Mordacci, R. (a cura di), *Modelli di medicina. Crisi e attualità dell'idea di professione*, pp. 35-59, Europa Scienze Umane ed., Milano.

Reich W. T., (ed.), (1995), *Encyclopedia of Bioethics*, 2nd edition, New York, Simon and Schuster MacMillan.

Reich W. T., Jecker N.S., (1995), *Care*, in Reich W. T., (ed.), (1995), *Encyclopedia of Bioethics*, 2nd edition, New York, Simon and Schuster MacMillan, pp. 319-344.

Reiter-Theil S., (2000), *Ethics consultation on demand: concepts, practical experiences and a case study*, in "Journal of Medical Ethics", 3, pp. 198-203.

Richardson H., (1990), *Specifying norms as a way to resolve concrete ethical problems*, in "Philosophy and Pubblic Affairs", 19, pp. 279-310.

Ricoeur P., (1986), *Le modèle du texte: l'action sensée considérée comme un texte*, in *Du texte à l'action. Essais d'herméutique II*, Coll. Points- Essais, Paris.

Ricoeur P., (1990), *Soi-même comme un autre*, Seuil, Paris.

Ricoeur P., (1996), *Le trois niveaux du jugement médical*, in "Ésprit", 12, pp. 21-33.

Riis P., (1994), *Medical ethics in the European Community*, in: "Journal of Medical Ethics", 1, pp. 7-12.

Rose G., (1992), *The strategy of preventive medicine*, Oxford University Press, Oxford

Rothman D., (1993), *Strangers at the bedside*, Basic Books, New York.

Roy D., (1987), *L'éthique clinique*, in *La responsabilité étique dans le developpement biomedical*, Cathalises, Louvain-La-Neuve, pp. 263-281.

Roy D.J., Williams J. R., Dickens B. M., Baudouin J. L., (1995), *La bioéthique, ses fondements et ses controverses*, ERPI, Montreal.

Sala R., (2003), *Bioetica e pluralismo dei valori. Tolleranza, principi, ideali morali*, Liguori, Napoli.

Sandel M., (1982), *Liberalism and the Limits of Justice*, Cambridge University Press, Cambridge.

Scheler M., (1916), *Der Formalismus in der Ethik un die materiale Wertethik. Neuer Versuch der Grndlegung eines ethischen Personalismus*, Niemeyer, Halle.

Sgreccia E., (2000), *Manuale di bioetica*, 2 voll., Vita e Pensiero, Milano.

Shapiro R. et al., (2000), *Managed care: effects on the physician-patient relationship*, in "Cambridge Quarterly of Health Care Ethics", 9, pp. 71-81.

Sherwin S., (1989), *Feminist and medical ethics: two different approaches to contextual ethics*, in "Hypathia", 2, pp. 57-72.

Sherwin S., (1992), *No longer patient: feminist ethics and health care*, Temple University Press, Philadelphia.

Siegler M., (1985), *The progression of medicine. From physician paternalism to patient autonomy to bureaucratic parsimony*, in "Arch. Intern. Med.", 145, pp. 716-719.

Siegler M., Pellegrino E. D., Singer P. A., (1990), *Clinical medical ethics*, in "The Journal of Clinical Ethics", 1, pp. 5-9.

Spencer T., (1997), *A new role of institutional ethics committees: organizational ethics*, in: "The Journal of Clinical ethics", 4, p. 372-376.

Stout J., (1988), *Ethics after Babel. The language of morals and their discontents,* Beacon Press, Boston.

Sulmasy D. P., (1993), *What's so special about medicine?,* in "Theoretical Medicine", 14, pp. 27-42.

Sumner L.W., Boyle J. J., (eds), (1996), *Philosophical perspectives on bioethics*, University of Toronto Press, Toronto.

Svenaeus F., (1999), *The hermeneutics of medicine and the phenomenology of health. Steps towards a philosophy of medical practice*, Linköping University.

Szumowski W., (1937), *L'histoire de la médecine et la réforme des études médicales*, Librairie Gebethner & Wolff, Cracovie.

Szumowski W., (1949), *La Philosophie de la Médecine, son histoire, son essence, sa dénomination et sa définition* in "Archives Internationales d'Histoire des Sciences" 2(9), pp. 1097-1139.

Taylor Ch., (1989), *Sources of the self. The making of modern identity,* Cambridge University Press, Cambridge.

The goals of medicine: setting new priorities, (1996), in "Hastings Center Report", 26, pp. S1-S27.

Thomasma D. C., (1994), *Clinical ethics as medical hermeneutics*, in "Theoretical Medicine", 15, pp. 93-111.

Thomasma D. C., (1997) *Bioethics and international human rights*, in "Journal of Law, Medicine & Ethics", 25, pp. 295-306.

Tollefsen C., (1995), *Foundationalism defended. Essays on epistemology, ethics and aesthetics,* Cambridge University Press, Bethesda.

Tongeren P. Van, (1994), *The relation of narrativity and hermeneutics to an adequate practical ethics,* in "Ethical Perspectives" 1(2): 57-70.

Toulmin S., (1981), *The tyranny of principles*, in "Hastings Center Report", 11, 6, pp. 31-39.

Toulmin S., (1982), *How medicine saved the life of ethics*, in "Perspectives in Biology and Medicine", 25, pp. 736-750.

Tronto J.C., (1993), *Moral boundaries. A political argument for an ethic of care.* Routledge, New York.

Tyminiecka A. T., Zalewski Z., (a cura di), (2000), *Life: the human being between life and death: a dialogue between medicine and philosophy*, in "Analecta Husserliana", Kluwer Academic Publishers, Dordrecht, LXIV, pp. 209-221.

Veatch R. M., (ed.), (1997), *Medical ethics*, Jones and Bartlett Publishers, Sudbury Mass., (2nd).

Veatch R., (1989), *Cross cultural perspectives in medical ethics,* Jones and Bartlett, Boston.

Veatch R.M., (2000), *The internal and external sources of morality for medicine*, in Thomasma D., Kissell J. L. (eds.), *The healthcare professional as friend and healer*, Georgetown University Press, Washington D.C., pp. 75-86.

Velleman D., *The possibility of practical reason*, in "Ethics", 106, pp. 694-727.

Verspieren P., (1997), *Une expérience de réflexion éthique sur la pratique clinique*, in "Le Supplement", 202, pp. 129-142.

Vetlesen A. J., (1994), *Perception, empathy and judgment. An inquiry into preconditions of moral performance*, Pennsylvania, Pennsylvania University Press.

Viafora C. (1995), *Comitati etici. Una proposta bioetica per il mondo sanitario*, Fondazione Lanza. Gregoriana Libreria Editrice.

Viafora C. (1999), *Toward a methodology for the ethical analysis of clinical practice,* in "Medicine, Health care and Philosophy, An European Journal", pp. 283-297

Viafora C. (2000), *Ethical function in hospital ethics committees: potentialities and ambiguites*, in: J. Glasa (ed), *Ethics Committees in central and eastern Europe*, Council of Europe, Strasbourg.

Viafora C., Gobber D., Drigo P., (2002), *Ethics committees in Italy: Theoretical framework and the experience of the bioethics committee of the department of pediatrics, University of Padua*, in Lebeer G., *Ethical function in hospital*, pp. 67-79.

Vineis P.,(1999), *Nel crepuscolo della probabilità. La medicina tra scienza ed etica*, Einaudi, Torino.

Vollmann J., (1995), *Der klinische ethik-ein konzept mit zukunft? Zur integration von philosophischer ethik in die praktische medizin*, in "Ethik in der Medizin", 7, pp.181-192.

Weir R., (1987), *Infants ethics committees. Ethical advisors or legal watchdogs?*, in: "Law Medicine and Health Care", 3, p. 99-109.

Weisz G., (ed.), (1990), *Social science perspectives on medical ethics,* Kluwer Academic Publishers, Dordrecht.

Welie J., (1998), *Clinical ethics: theory or practice?,* in "Theoretical Medicine and Bioethics", 19, pp. 295-312.

Welie J., (1995), *La centralità della funzione formativa nell'esperienza dei Comitati etici. Riflessioni del Segretario di un Comitato etico europeo*, in C. Viafora (a cura di), *Comitati etici*, pp. 75-90.

Welie J., (1998), *In the face of suffering. The philosophical-anthropological foundations of clinical ethics,* Creighton University Press, Omaha.

Wenger N., (2000), *The healthcare ethics committee model of the future builded on deficiencies of the past*, in "HEC forum", 12, pp. 33-38.

Widdershoven G., (2002), *Alternatives to principlism: phenomenology, deconstruction, hermeneutics*, in Dickenson D., Fulbord B., Murray T. (eds): *Healthcare ethics and human values. An introductory text with readings and case studies*, Blackwell, Oxford, pp. 41-48.

Williams B., (1988), *Consequentialism and integrity*, in, Scheffler S., (ed.), *Consequentialism and its critics.* Oxford, pp. 20-50.

Wilson Ross J., (1986), *Handbook for hospital ethics committees*, American Hospital Association Publishing, Chicago.

Winkler E., (1996), *Philosophy and bioethics: contextualism versus the paradigm theory*, in Sumner L.W., Boyle J. J., (eds), *Philosophical perspectives on bioethics*, pp. 50-78.

Winkler E.R., (1998), *Applied ethics, overview*, in Chadwick, R. (ed.), *Encyclopedia of applied ethics*, Academic Press, San Diego, 1, pp. 191-196.

Wittgenstein L., (1974), *Philosophical Investigations,* 2nd edition, Oxford, Blackwell.

Wulff H. R., Pedersen S. A., Rosenberg R., (1986), *Philosophy of medicine*, Blackwell Scientific Publications, London.

Wulff H.R., (1988), *Contemporary trends in healthcare ethics*, in Have, H.A.M.J. ten and Sass, H.-M. (eds.), *Consensus formation in healthcare ethics,* Kluwer Academic Publishers, Dordrecht, pp. 63-72.

Zaner R. M., (1988), *Ethics and the clinical encounter*, Prentice Hall, Englewood Cliffs, New Jersey.

Zatti P., (1997), *Il principio di correttezza nel rapporto medico - paziente,* in Atti del Convegno "Il consenso informato", Centro Nazionale di prevenzione e di Difesa sociale - Fondazione dell'Istituto Europeo di Oncologia, Milano.

Zatti P., (2004), *Spunti in tema di libertà di cura: tra sperimentazione e terapia innovativa*, in S*tudi in onore di P.Schlesinger,* Giuffrè, Milamo.

Zwart H., (1998), *Medicine, symbolization and the "real" body - Lacan's understanding of medical science,* in "Medicine, Health Care and Philosophy" 1(2): 107-117.

International Library of Ethics, Law, and the New Medicine

1. L. Nordenfelt: *Action, Ability and Health.* Essays in the Philosophy of Action and Welfare. 2000 ISBN 0-7923-6206-3
2. J. Bergsma and D.C. Thomasma: *Autonomy and Clinical Medicine.* Renewing the Health Professional Relation with the Patient. 2000 ISBN 0-7923-6207-1
3. S. Rinken: *The AIDS Crisis and the Modern Self.* Biographical Self-Construction in the Awareness of Finitude. 2000 ISBN 0-7923-6371-X
4. M. Verweij: *Preventive Medicine Between Obligation and Aspiration.* 2000 ISBN 0-7923-6691-3
5. F. Svenaeus: *The Hermeneutics of Medicine and the Phenomenology of Health.* Steps Towards a Philosophy of Medical Practice. 2001 ISBN 0-7923-6757-X
6. D.M. Vukadinovich and S.L. Krinsky: *Ethics and Law in Modern Medicine.* Hypothetical Case Studies. 2001 ISBN 1-4020-0088-X
7. D.C. Thomasma, D.N. Weisstub and C. Hervé (eds.): *Personhood and Health Care.* 2001 ISBN 1-4020-0098-7
8. H. ten Have and B. Gordijn (eds.): *Bioethics in a European Perspective.* 2001 ISBN 1-4020-0126-6
9. P.-A. Tengland: *Mental Health.* A Philosophical Analysis. 2001 ISBN 1-4020-0179-7
10. D.N. Weisstub, D.C. Thomasma, S. Gauthier and G.F. Tomossy (eds.) : *Aging: Culture, Health, and Social Change.* 2001 ISBN 1-4020-0180-0
11. D.N. Weisstub, D.C. Thomasma, S. Gauthier and G.F. Tomossy (eds.) : *Aging: Caring for our Elders.* 2001 ISBN 1-4020-0181-9
12. D.N. Weisstub, D.C. Thomasma, S. Gauthier and G.F. Tomossy (eds.) : *Aging: Decisions at the End of Life.* 2001 ISBN 1-4020-0182-7
(Set ISBN for Vols. 10-12: 1-4020-0183-5)
13. M.J. Commers: *Determinants of Health: Theory, Understanding, Portrayal, Policy.* 2002 ISBN 1-4020-0809-0
14. I.N. Olver: *Is Death Ever Preferable to Life?* 2002 ISBN 1-4020-1029-X
15. C. Kopp: *The New Era of AIDS.* HIV and Medicine in Times of Transition. 2003 ISBN 1-4020-1048-6
16. R.L. Sturman: *Six Lives in Jerusalem.* End-of-Life Decisions in Jerusalem - Cultural, Medical, Ethical and Legal Considerations. 2003 ISBN 1-4020-1725-1
17. D.C. Wertz and J.C. Fletcher: *Genetics and Ethics in Global Perspective.* 2004 ISBN 1-4020-1768-5
18. J.B.R. Gaie: *The Ethics of Medical Involvement in Capital Punishment.* A Philosophical Discussion. 2004 ISBN 1-4020-1764-2
19. M. Boylan (ed.): *Public Health Policy and Ethics.* 2004 ISBN 1-4020-1762-6; Pb 1-4020-1763-4
20. R. Cohen-Almagor: *Euthanasia in the Netherlands.* The Policy and Practice of Mercy Killing. 2004 ISBN 1-4020-2250-6
21. D.C. Thomasma and D.N. Weisstub (eds.): *The Variables of Moral Capacity.* 2004 ISBN 1-4020-2551-3
22. D.R. Waring: *Medical Benefit and the Human Lottery.* An Egalitarian Approach. 2004 ISBN 1-4020-2970-5
23. P. McCullagh: *Conscious in a Vegetative State? A Critique of the PVS Concept.* 2004 ISBN 1-4020-2629-3

International Library of Ethics, Law, and the New Medicine

24. L. Romanucci-Ross and L.R. Tancredi: *When Law and Medicine Meet: A Cultural View*. 2004 ISBN 1-4020-2756-7
25. G.P. Smith II: *The Christian Religion and Biotechnology*. A Search for Principled Decision-making. 2005 ISBN 1-4020-3146-7
26. C. Viafora (ed.): *Clinical Bioethics*. A Search for the Foundations. 2005 ISBN 1-4020-3592-6

MIX
Papier aus verantwortungsvollen Quellen
Paper from responsible sources
FSC® C105338
www.fsc.org

If you have any concerns about our products,
you can contact us on
ProductSafety@springernature.com

In case Publisher is established outside the EU,
the EU authorized representative is:
Springer Nature Customer Service Center GmbH
Europaplatz 3, 69115 Heidelberg, Germany

Printed by Libri Plureos GmbH
in Hamburg, Germany